MW01529387

PHOTOGRAPHIC GUIDE TO THE
MARINE MAMMALS OF THE NORTH ATLANTIC

Other titles in the *Photographic Guides* series

Photographic Guide to Sea and Shore Life of Britain and North-West Europe
Ray Gibson, Benedict Hextall, and Alex Rogers

Photographic Guide to the Butterflies of Britain and Europe
Tom Tolman

Photographic Guide to Minerals of the World
Ole Johnsen

PHOTOGRAPHIC GUIDE TO THE
MARINE MAMMALS OF
THE NORTH ATLANTIC

Carl Christian Kinze

ILLUSTRATED BY
Birgitte Rubæk

TRANSLATED BY
David A. Christie

OXFORD
UNIVERSITY PRESS

OXFORD
UNIVERSITY PRESS

Great Clarendon Street, Oxford OX2 6DP

Oxford University Press is a department of the University of Oxford.
It furthers the University's objective of excellence in research, scholarship,
and education by publishing worldwide in

Oxford   New York
Auckland   Bangkok   Buenos Aires   Cape Town   Chennai
Dar es Salaam   Delhi   Hong Kong   Istanbul   Karachi   Kolkata
Kuala Lumpur   Madrid   Melbourne   Mexico City   Mumbai   Nairobi
Sao Paulo   Shanghai   Singapore   Taipei   Tokyo   Toronto

and an associated company in Berlin Ibadan

Oxford is a registered trade mark of Oxford University Press
in the UK and in certain other countries

Published in the United States and Canada by Princeton University Press,
41 William Street, Princeton, New Jersey 08540

Originally published under the title: *Havpattedyr i Nordatlanten*
© G.E.C. Gads Forlag. Aktieselskabet af 1994, Denmark 2001

English edition © Princeton University Press 2002

The moral rights of the author have been asserted

Database right Oxford University Press (maker)

First published 2002

British Library Cataloguing in Publication Data

Data available

Library of Congress Cataloging in Publication Data

ISBN 0 19 852625 3

10 9 8 7 6 5 4 3 2 1

Typeset by D & N Publishing, Marlborough, Wiltshire
Printed by Narayana Press, Denmark

Contents

Baleen whales Mysticeti *119*

Seals and walruses *147*

Other marine-mammal groups *173*

Preface

In recent years, increasing numbers of people have become more aware of marine mammals. Many have made the leap from a passive enjoyment of long-distance observation to a 'personal experience', or even to an active involvement in matters concerning marine mammals. Once a person's enthusiasm has been aroused, he or she will wish not only to watch marine mammals but also to know more about them. Which species are involved? Why are they at a particular place at a particular time?

To be able to expand one's knowledge, i.e. to identify species or describe their behaviour, certain preparatory steps are necessary. The enthusiast has to become acquainted with new terms and basic concepts, but, once this is achieved, one's experience of nature acquires an extra dimension.

For many of the species of marine mammal found in the North Atlantic there are quite big gaps in our knowledge, and in some cases even the most basic information is lacking. This, however, increases the value of observations and makes the challenge a greater one. 'Professional' marine-mammal researchers cannot be everywhere. An 'amateur' who happens to be in the right place at the right time can be the first person to 'discover' entirely new information on the distribution, migrations and behaviour of individual species. Ultimately, it is this knowledge that provides the foundation for a rational management and, if necessary, protection of marine-mammal populations.

During the preparation of this book, I have drawn on the help of a large number of colleagues. Many have provided direct assistance; others have helped indirectly through the numerous instances of knowledge-sharing which they have provided in the literature. There are too many individuals to mention by name, but none is forgotten: I wish here to thank all of them.

This book does not provide answers to all conceivable and inconceivable questions that could be asked about marine mammals, nor does it deal with attitudes regarding cetaceans and politics. On the other hand, I hope that it will inspire the reader to go out and observe marine mammals in the North Atlantic and its adjoining waters—both at sea and along the coasts.

No book is perfect. It is almost certain that new information on distribution, behaviour and other aspects will come to light. I therefore urgently request all readers to send in new information and to point out any errors, so that these can be corrected in a future edition. These should be sent to Gads Forlag.

Frederiksberg, June 2002
Carl Christian Kinze

Structure of the book and how to use it

The structure of the book and how to use it are described below. The following chapter deals with the North Atlantic as a habitat. There is then a brief definition of the term 'marine mammals', followed by separate chapters on cetaceans (whales, dolphins and porpoises), pinnipeds (seals and walruses), manatees, and marine carnivores, with descriptions of the 53 marine mammals which occur in the North Atlantic. These comprise 40 cetaceans, 8 pinnipeds, 2 manatee species, Polar Bear, and the 2 species of coastal otters found on each respective side of the North Atlantic.

Limits of the sea area covered by the book

This book covers the North Atlantic north of 25ºN, i.e. the waters from the North Pole south to Africa's west coast, including the Canary Islands, in the east, and down the east coast of North America to the

The North Atlantic as defined in this book.

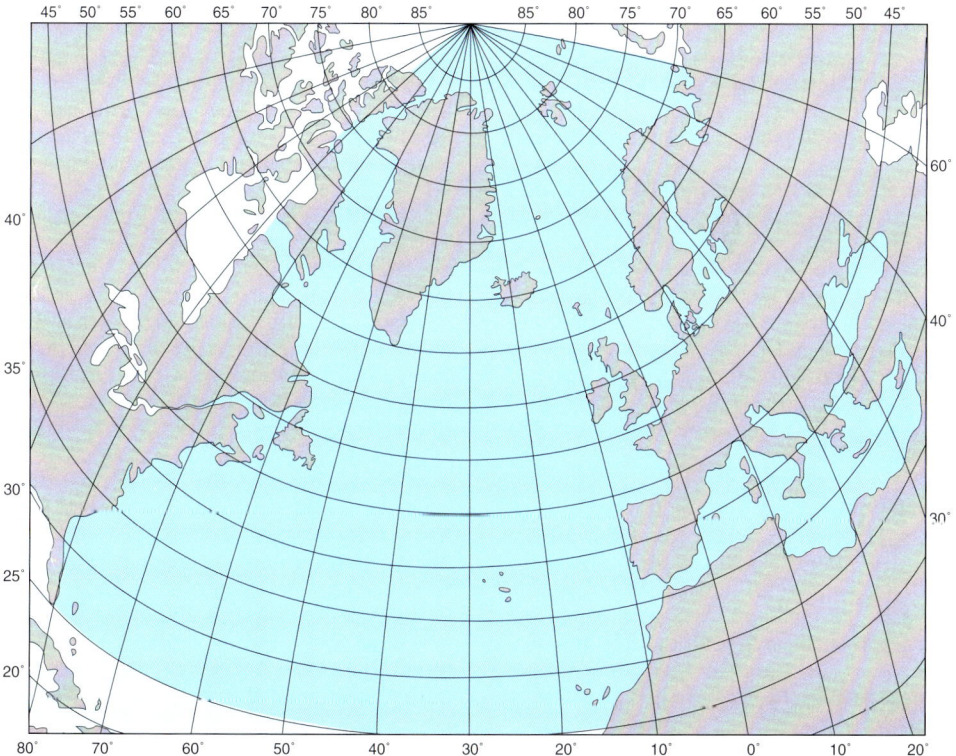

southern tip of Florida in the west. This area includes, among others, the Baltic, Mediterranean and Black Seas, the Gulf of Saint Lawrence, Hudson Strait and numerous other, smaller seas (*see* map).

General information

The individual marine-mammal groups are described separately. For each group, the species descriptions are preceded by a presentation of the particular terms and characteristics that are used in the identification and description of the species.

Species descriptions

For each species, the following information is given, where known:

Name: English name(s), scientific name, and names in the three languages French, German and Spanish.

Description: Details on the species' body shape, the lengths and weights of adult and of newborn young, number of teeth/baleen plates, colour pattern and variations within the species (differences between sexes, young and old individuals, and any differences between geographical populations).

Coloration: A typical full-grown individual is shown. For species with substantial sexual or age-dependent variations, both sexes or adult and young are presented.

Photographs: Show the species as it is seen in the field.

Behaviour: Details of group size and of any group structure, if these are known.
   Also mentioned here are migrations, diving behaviour, and behaviour at the surface or on land. Special behavioural patterns associated with mating or foraging may, however, be mentioned under 'Breeding' or 'Food'.

Breeding: The age and/or length at sexual maturity are given for both sexes, so far as these are known. Also mentioned here are the gestation period (duration of pregnancy) and the lactation period (suckling period), the average number of young per year, times of mating and of births, and any special behaviours associated with breeding.

Food: The most important food items are mentioned, as are particular patterns of behaviour associated with foraging.

Natural enemies: Natural enemies, either documented or presumed, are indicated.

Relationship with humans: Reference is made to any commercial or local whaling, as well as accidental by-catches, and whether any animals are known to have been kept in captivity.

Similar species: Mentioned under this heading are any similar species which can be confused with the one described, and how they can be distinguished from each other. If relevant, reference is made to a comparative plate or table.

Distribution maps

On these maps, the species' known and presumed range, i.e. where it is likely to be observed with reasonable probability, is marked in colour. In the process of identifying a marine mammal, one can use the maps to eliminate species which do not normally occur in the area (geographical exclusion), but this method should not be used on its own. Many species may be seen as stragglers outside their normal or known range. Where there is a certain regularity in the routes taken by migrants, these are indicated by arrows on the map.
   The accompanying caption begins with a general summary of the species' preferred living area. Its geographical occurrence is then treated in more detail.

Comparative plates

In several places in the book, comparative plates are included in order to facilitate identification in the field:

Dolphins page 42
'Blackfish' page 73
Seals page 148

How to proceed

When a marine mammal is sighted, one must first determine to which of the book's main groups it belongs. This may be done by using the identification key below. Then you will need to use the illustrations, comparative plates and text in order to proceed to the species' name.

Identification key

1. – Without or with very little fur . . . . go to 2
   – With short or well-developed fur . . . . go to 3

2. – With two limbs and forked tail fin . . . . Cetaceans, go to 4
   – With two limbs and rounded tail fin . . . . **Manatees**, pages 174–177

3. – Fur short, fore and hind limbs very different . . . . Seals, go to 11
   – Fur dense, fore and hind limbs not very different . . . .**Polar Bear**, page 178; **Otters**, page 180

Cetaceans

4. – With teeth and single blowhole . . . . Toothed whales, go to 5
   – With baleen and double blowhole . . . . Baleen whales, go to 10

5. – With dorsal fin . . . . .go to 6
   – Without dorsal fin . . . . **Beluga and Narwhal**, pages 94–98

6. – Tail notch . . . . go to 7
   – No tail notch . . . . **Beaked whales**, pages 99–111

7. – Head streamlined . . . . go to 8
   – Head box-shaped . . . . Sperm whales, go to 9

8. – Teeth conical . . . . **Dolphins**, pages 41–90
   – Teeth spatulate . . . . **Harbour Porpoise**, pages 91–93

9. – Blowhole at front of head and on left, large . . . . **Sperm Whale**, pages 112–114
   – Blowhole on top of head, small . . . . **Dwarf and Pygmy Sperm Whales**, pages 115–118

10. – Without dorsal fin and throat grooves, baleen plates long . . . . . **Right whales**, pages 119–123
    – With dorsal fin and throat grooves, baleen plates short . . . . **Rorquals**, pages 124–145

Seals

11. – Without tusks, backward-directed hind limbs . . . . **Seals**, pages 146–168
    – With tusks, hind limbs supporting body . . . . **Walruses**, pages 169–171

The North Atlantic

As mentioned earlier, the North Atlantic is the Atlantic Ocean north of 25°N. This sea area covers arctic, subarctic, temperate and subtropical regions and excludes only the purely tropical areas.

There is a wealth of information available on physical and chemical conditions in these waters, so only a brief account will be given here of the conditions which are of immediate importance for an understanding of the distribution and ecology of marine mammals: sea depth, sea temperature, ice conditions, and the salinity of the water.

Fluctuations in temperature and salinity cause movements of the water masses in the form of ocean currents. Where water masses of different temperature and salinity meet, thermoclines and haloclines, or fronts, are created. Elsewhere, the bottom water is forced up to the surface and creates what is known as 'upwelling'; this is seen especially where deep currents meet land. Warm air over the ocean results in evaporation, causing the salinity of the sea to rise, while cold air over the ocean can lead to ice formation. All these factors also have an influence on the distribution of marine mammals.

Water depth

The sea can be subdivided according to its depth into four zones, which, as habitats, differ just as much as do e.g. savanna and rainforest on dry land.

The first zone, the *tidal zone*, is exposed at high water, but is covered by the high tide. This zone is seldom exploited by cetaceans, but is used extensively by various seal species. The next zone consists of the shallow-water coastal areas, the *shelf areas (see* figure), which begin directly off the tidal zone and slope down to a depth of 200 m. A large number of seal species live here, as do the so-called coastal cetacean species. The third zone is represented by the *sea above the continental slope*, where, over a relatively short horizontal distance, the sea bottom plummets to the ocean depths. The slope is often characterised by a great abundance of fish, and therefore attracts a number of marine mammals. The *deep sea* of the open ocean makes up the fourth zone. This is where the oceanic species live. The deep sea contains both 'ridges' or rises, and large flat 'valleys' known as basins, as well as even deeper trenches. The marine mammals in this zone, especially the whales, lead a pelagic life, i.e. they live in, and find their food in, the open waters, but some species also dive to the bottom for food, descending to depths of several thousand metres.

North Atlantic shelf areas

In the North Atlantic, the shelf areas are not uniformly distributed. The largest shelf area is found in the eastern part of the North Atlantic, in Europe including Bay of Biscay, the Irish Sea, the English Channel and the North Sea, whereas the shelf areas in the western North Atlantic are far smaller in extent. This, of course, has consequences for the distribution of marine mammals. Not surprisingly, for

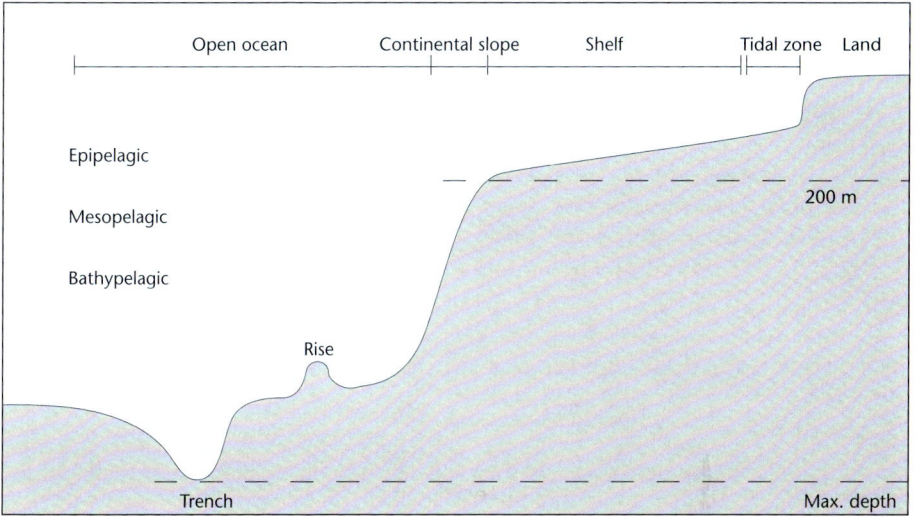

Divisions of the ocean according to depth.

Distribution of shelves.

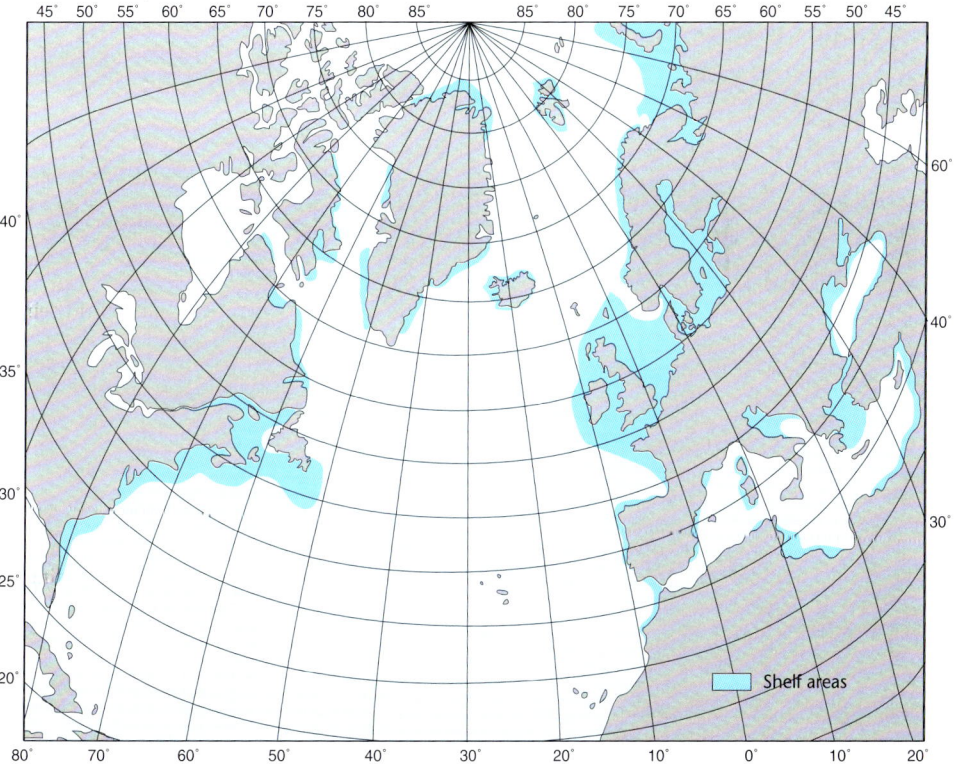

instance, the largest populations of Harbour Porpoises are found in the eastern shelf area. These circumstances perhaps also explain why the shelf species the White-beaked Dolphin (*Lagenorhynchus albirostris*) is much more common on the eastern side of the North Atlantic than on the western side, whereas the very opposite applies to the oceanic White-sided Dolphin (*Lagenorhynchus acutus*).

Oceanic conditions close to land

In some places the shelf rim is very narrow, with relatively deep water in close vicinity to land. Certain oceanic islands, e.g. the Azores and the Canary Islands, lie in the middle of the open sea and surrounded only by a narrow coastal rim.

Here, therefore, far more oceanic species can be observed just a few kilometres from land than is possible in, for example, the North Sea. In some places deep channels are cut into the shelf sea, e.g. The Gully off the Nova Scotian coast and the Norwegian Channel in the northern North Sea. These channels attract oceanic species, but at times with fatal consequences when the animals lose their way in the adjacent shallow sea areas.

Temperature and salinity

As already mentioned, sea areas are characterised also by temperature and salinity, which are subject to wide fluctuations as

Simplified diagram showing salinity in North Atlantic surface waters.

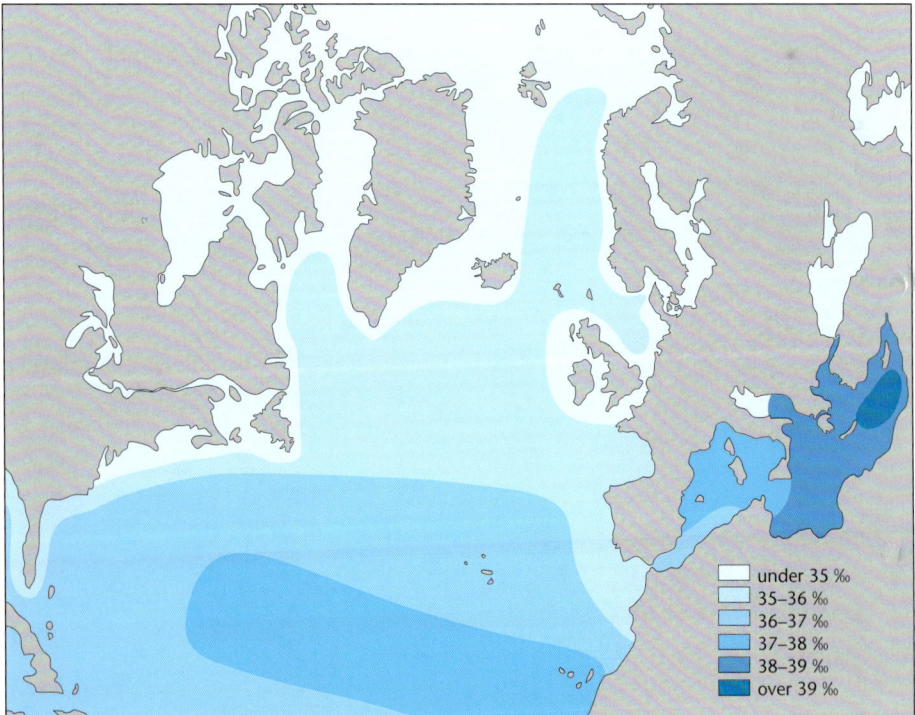

Legend:
- under 35 ‰
- 35–36 ‰
- 36–37 ‰
- 37–38 ‰
- 38–39 ‰
- over 39 ‰

the water masses are in constant motion. The wind creates ocean currents which can carry water masses over large distances. When cold and warm water masses meet, a gradual mixing occurs. This mixing is not, however, complete, because cold water, which is typically more saline, is heavier and is therefore forced down beneath water masses which are lighter as a result of their higher temperature and/or lower salinity. In some places a very sharp boundary is formed between the water masses, known as a *front*. In others special areas of *upwelling* occur, i.e. areas where deep water is forced up to the surface. As the upwelling water is very rich in nutrients, concentrations of prey animals occur here and these, of course, will attract predators: the marine mammals.

Temperature zones in the North Atlantic

From north to south in the North Atlantic, there is first a belt of arctic water, then a belt of temperate water and, in the south, belts of subtropical and tropical water (*see* figure below). Ocean currents and the fluctuations in their strength mean that displacements between these principal temperature belts take place more or less frequently. Because of the Gulf Stream, the temperate sea areas extend farther north on the eastern side of the North Atlantic than on the western side. The western North Atlantic is, by contrast, under the influence of cold ocean currents, so that (mixed) arctic water reaches farther south here than on the eastern side. This, of

Temperature zones in the North Atlantic on basis of the average surface-water temperature in February. Arctic sea areas below 0°C, temperate seas 5-10°C, subtropical seas 10-20°C, and tropical seas above 20°C.

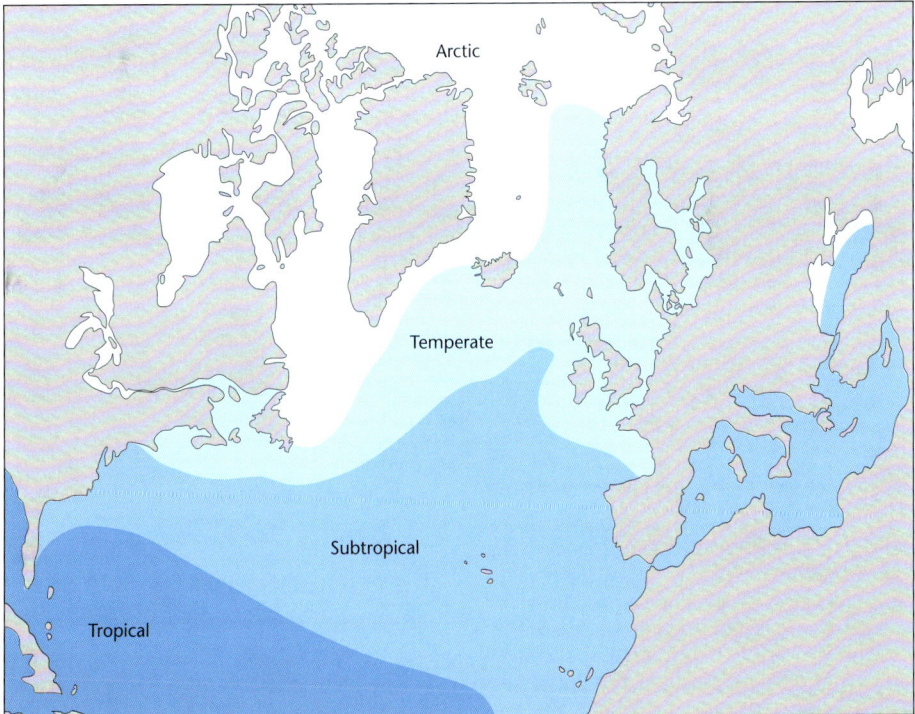

Areas in the North Atlantic where ice forms.

course, is reflected in the distribution and occurrence of marine mammals: species associated with cold water occur farther south on the western side than in the east, where, on the other hand, species associated with warmer water occur much farther north than they do in the west.

Ice

In the northernmost part of the North Atlantic, and in the Baltic Sea, the northern Black Sea and the Sea of Azov, ice is regularly present. This creates problems for some species, which may move seasonally to ice-free areas. This applies, for example, to porpoise populations in the Baltic and Black Seas.

For other species, the ice is, on the contrary, a part of their life. Several seal species breed on the ice and are able to keep breathing holes open throughout the winter.

For yet others, the situation is a mixed one. Certain whale species associated with the edge of the ice can, for instance, become entrapped in holes in the ice, and will perish when these close up. This phenomenon is known as *sasat* in Greenland.

The Polar Bear is at home walking on drift ice and on solid ice.

Ocean currents

The North Atlantic is influenced by big ocean currents, first and foremost the *Gulf Stream*. As its name indicates, this originates in the Gulf of Mexico, and it is therefore relatively warm and salty. The Gulf Stream runs from its source northwards along the east coast of North America until forced eastwards by the cold Labrador Current.

Halfway to Europe, the *Canaries Current* separates off to the south and the *Irminger Current* to the north. The Canaries Current gradually becomes mixed and therefore relatively cooled on its way southwards; its water is therefore cold in comparison with that which it meets on its way down along the West African coast. An area of upwelling is formed here and produces an abundance of fish, which attracts marine mammals. The Irminger Current runs up to Greenland and then down its east coast, rounding Cape Farvel and heading northwards.

The Gulf Stream continues along the west coast of the British Isles and splits into two arms. One enters the North Sea, while the other runs along the Norwegian coast into the Barents Sea. A side branch of the latter swings northwest to Svalbard, where it becomes an undercurrent beneath the comparatively fresh and cold arctic waters (not shown on map).

Geographical divisions

Lengthwise, the North Atlantic is divided by the *Mid-Atlantic Ridge* into an eastern and a western half. In many places, as at the Azores and Iceland, this ridge is marked by the formation of volcanic islands. On each side of the ridge are deep-sea areas, known as basins. The basins are sometimes further divided by submarine plateaux or rises.

Ocean currents in the North Atlantic.

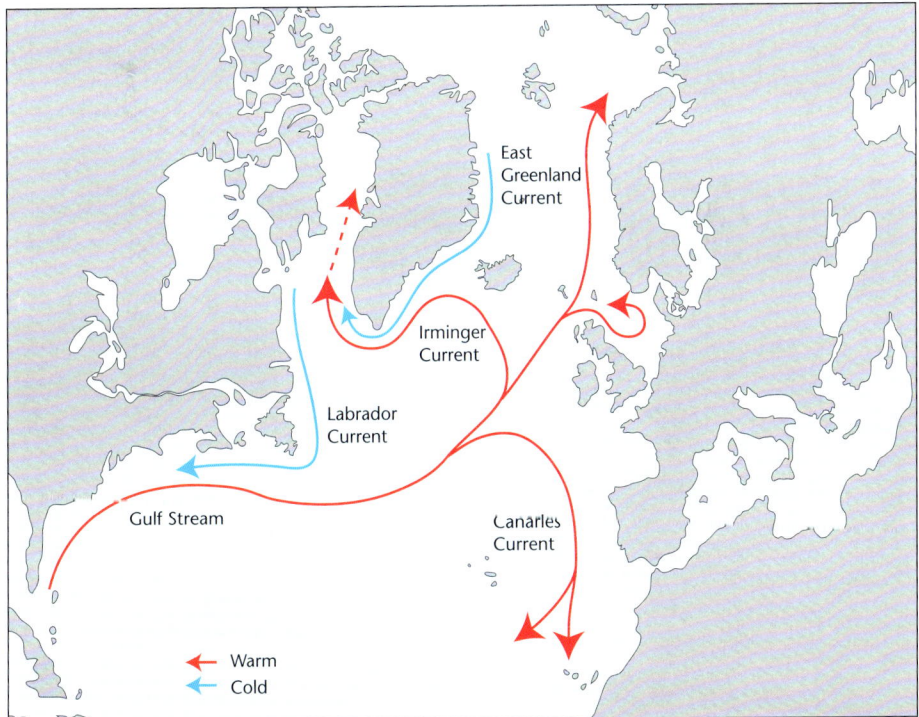

East Greenland Current

Irminger Current

Labrador Current

Gulf Stream

Canaries Current

Warm
Cold

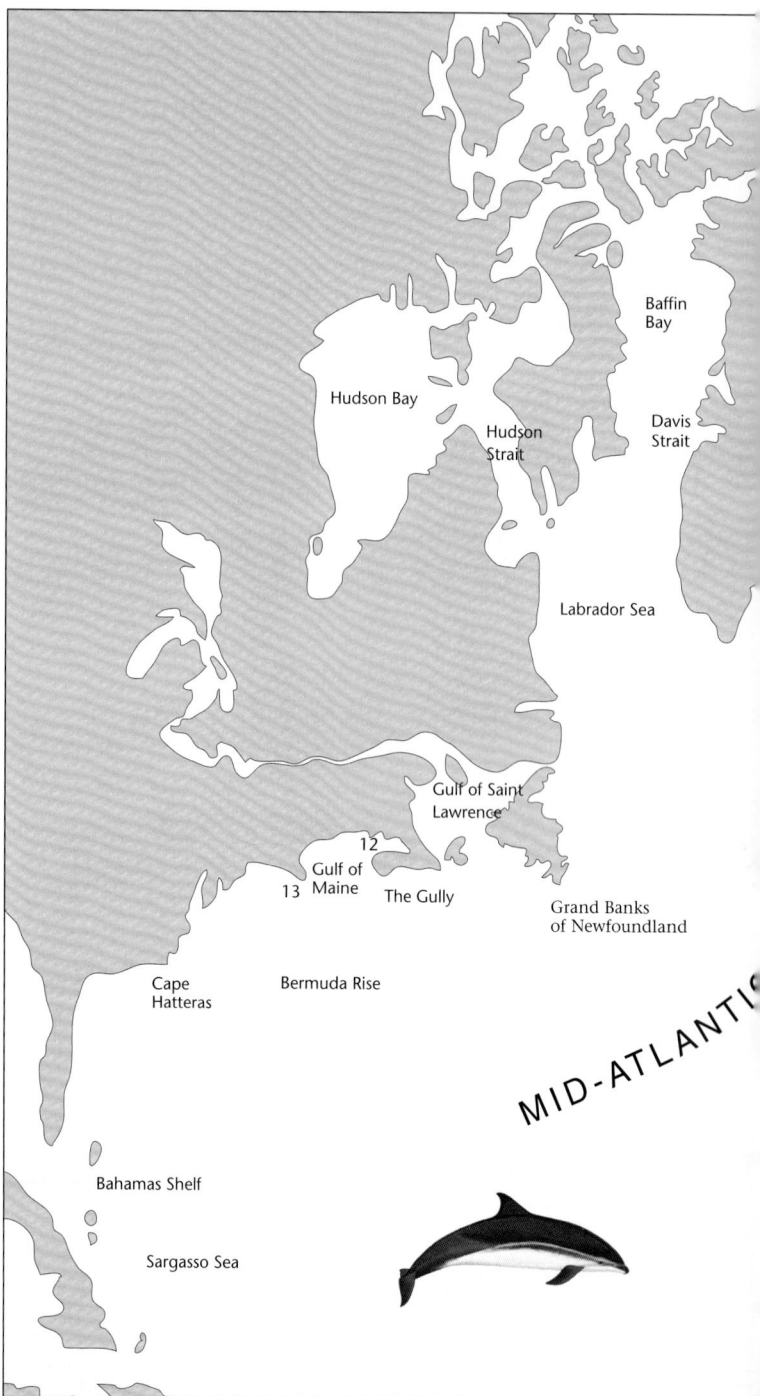

Baffin
Bay

Hudson Bay

Davis
Strait

Hudson
Strait

Labrador Sea

Gulf of Saint
Lawrence

12

Gulf of
13 Maine   The Gully

Grand Banks
of Newfoundland

Cape
Hatteras

Bermuda Rise

MID-ATLANTI

Bahamas Shelf

Sargasso Sea

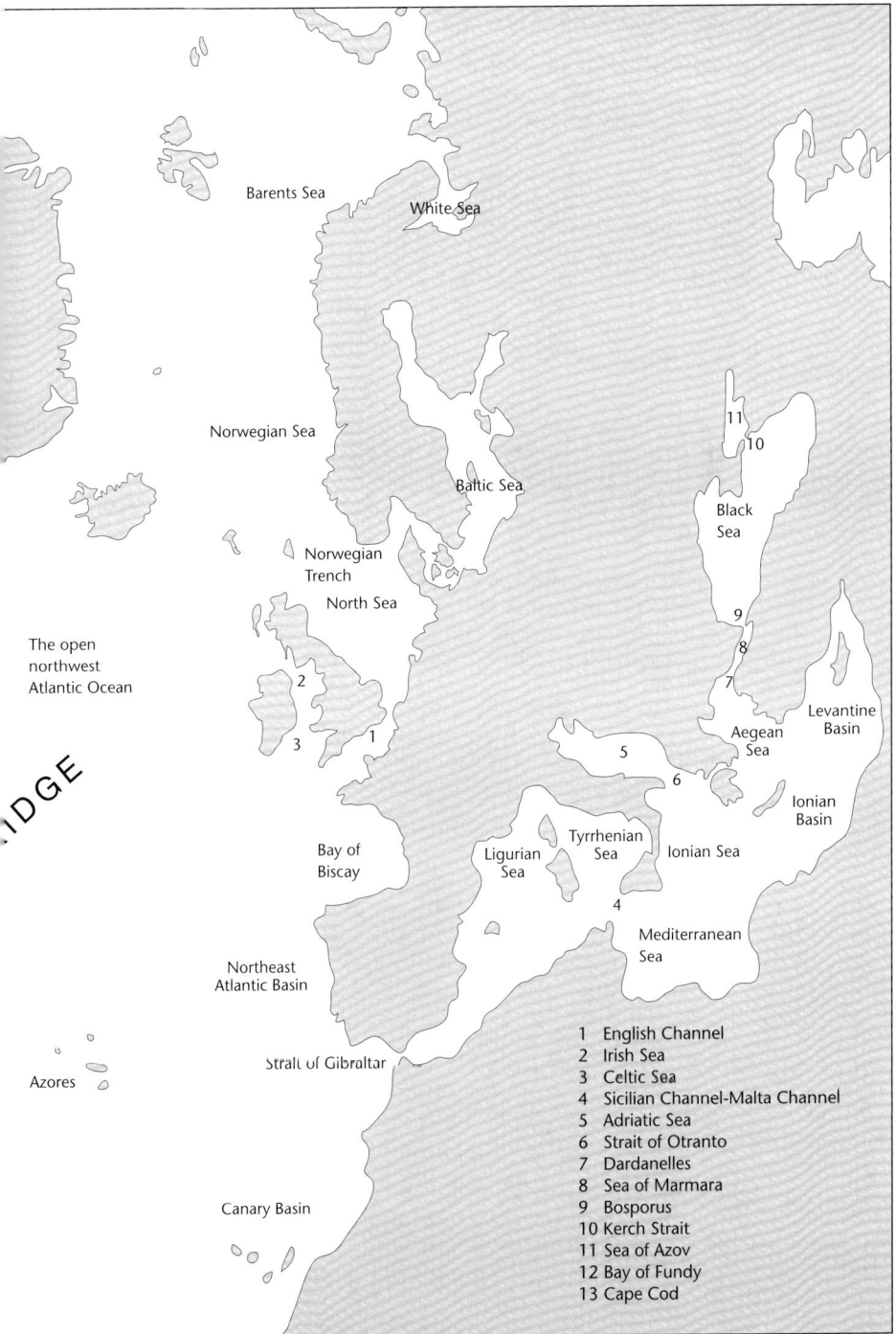

Barents Sea

White Sea

Norwegian Sea

Baltic Sea

Black Sea

11

10

9

8

7

Norwegian Trench

North Sea

The open northwest Atlantic Ocean

2

3

1

5

6

Aegean Sea

Levantine Basin

Ionian Basin

IDGE

Bay of Biscay

Ligurian Sea

Tyrrhenian Sea

Ionian Sea

4

Mediterranean Sea

Northeast Atlantic Basin

Strait of Gibraltar

Azores

Canary Basin

1 English Channel
2 Irish Sea
3 Celtic Sea
4 Sicilian Channel-Malta Channel
5 Adriatic Sea
6 Strait of Otranto
7 Dardanelles
8 Sea of Marmara
9 Bosporus
10 Kerch Strait
11 Sea of Azov
12 Bay of Fundy
13 Cape Cod

The Norwegian Sea, the Barents Sea and the White Sea

The sea area bounded by northern east Greenland, Svalbard and Norway is called the *Norwegian Sea* and includes several basins separated by ridges, such as the Norwegian Basin with a maximum depth of almost 4,000 m. The Norwegian Sea is connected to the Arctic Ocean via the deep Fram Strait. To the east, it passes into a shelf area consisting of the *Barents Sea* and the *White Sea*. To the south, the Norwegian Trench leads into the *North Sea* and the *Baltic Sea*.

North Sea and Baltic Sea

The North Sea is a shelf sea. The southern part is rather shallow, with several flat banks with a water depth of only 15 m. The northern part is deeper and slopes down to the continental slope. In the northwest, there is a comparatively deep plateau at a depth of about 100 m. To the east, the Norwegian Trench extends as a finger right into the Skagerrak, thereby carrying oceanic deep water into the North Sea and the Baltic. In the southern part of the North Sea, tidal variations have resulted in the creation of extensive areas of tidal flats, the Wadden Sea being the largest.

In the North Sea, the greatest exchange of water takes place with the open Norwegian Sea, and this can involve influences from both warmer, oceanic water masses originating in the Gulf Stream and colder water masses. In the northern North Sea, fronts are often created where water masses of different temperature and salinity meet.

The North Sea's water exchange through the English Channel is much smaller, but from here an influx of warmer water can occur which can lead to the occurrence of sea mammals which normally live farther south.

The *Baltic Sea* is almost a freshwater sea, and is regarded by some scientists as one big river mouth. In winter, the northern and central Baltic is more or less extensively covered with ice. On the Polish coast, particular current conditions maintain an ice-free area, which in harsh winters can serve as a refuge for marine mammals, e.g. some of the Baltic's porpoises. Through the Kattegat and the Danish straits an exchange of water takes place between the Baltic and the North Sea. Fresh water moves out of the area—at the surface—and more saline water moves in at the bottom. In some years the salt water reaches farther into the Baltic than normal, and during such periods unusual marine mammals may occur.

From the English Channel to the West African coast

The North Sea is connected by the *English Channel* to a larger shelf area, which consists of the *Irish Sea* between Ireland and Britain and the *Celtic Sea* south of Ireland together with the shallow northeastern part of *Bay of Biscay*. In the sea area south of Iceland and west of Britain lies the *Northeastern Atlantic Basin*, a tongue-shaped area which comprises the southern and western parts of Biscay and to the south merges with the *Canary Basin*. The coastal shelf here is very narrow. In the basins, however, there are a number of islands and shoals of volcanic origin: Madeira and the Canary Islands. The deepest spot here is 6,700 m.

The Mediterranean Sea

Separated from the Atlantic itself by the barely 15-km-wide Strait of Gibraltar, the Mediterranean Sea is the most important of the North Atlantic's subordinate seas. Across the Strait of Gibraltar's sill of about 400 m a slow exchange of water takes place with the North Atlantic. A tongue of more saline (because of evaporation) water flows as an undercurrent out of the Mediterranean, and a surface stream of less saline water flows in. Cetaceans accompany these water masses into and out of the Mediterranean, and the Strait of Gibraltar is one of the best localities for whale-watching, with up to 14 cetacean species in one day.

The Mediterranean Sea is divided into a western and an eastern end, which meet at the *Sicilian Channel-Malta Channel* between Sicily and Tunisia. Some sub-areas have their own name. The *Tyrrhenian Sea* is bounded by the islands of Sicily, Sardinia and Corsica and the Italian mainland, and the sea here reaches depths of over 3,700 m. The waters between north Corsica and the French and Italian Riviera are known as the *Ligurian Sea*. In the eastern Mediterranean there are in addition the comparatively shallow *Adriatic Sea*, which is almost closed off like a large bay at the *Strait of Otranto*, and the deep *Ionian Sea*. Two more deep-sea basins are found here, the *Ionian Basin* and the *Levantine Basin*. The area of water between Greece and Turkey is called the *Aegean Sea* and is characterised by rather shallow water.

The Black Sea and secondary seas

The entrance to the Black Sea consists of the *Dardanelles*, the *Sea of Marmara* and the *Bosporus*. Through these straits a slow exchange of water with the Mediterranean takes place. A surface stream of rather fresh water flows out, and an undercurrent of more saline water flows into the Black Sea. The Black Sea is linked by the *Kerch Strait* to the *Sea of Azov*. Because of inflow from primarily the Danube, the Black Sea is fairly fresh. The northern third is rather shallow, whereas the rest consists of a deep 'bowl' which reaches 2,000 m at the bottom. The Black Sea, over a comparatively short distance, spans near-arctic conditions in the north to subtropical conditions along the Turkish coast. In the vertical plane it is divided into two sharply separated strata: an upper level with rich fauna and a lower level typified by hydrogen sulphide, where higher organisms are completely absent.

Baffin Bay, Davis Strait and the Labrador Sea

The northernmost part of the western North Atlantic is distinctly arctic. The sea between west Greenland and Baffin Island is called *Baffin Bay* and is connected by the *Davis Strait* to the *Labrador Sea*. From here there are links to the arctic waters of *Hudson Strait* and *Hudson Bay* (not covered by this book).

Gulf of Saint Lawrence, Bay of Fundy and Gulf of Maine

The largest shelf area in the western North Atlantic is the Gulf of Saint Lawrence, which to the east joins with the *Grand Banks of Newfoundland* and to the south with the *Gulf of Maine* and its northernmost outpost the *Bay of Fundy*.

The Gulf of Saint Lawrence is, like the Baltic Sea, a huge river mouth, influenced by substantial freshwater outflow from the Great Lakes. The very special conditions here have resulted in the establishment of a quite unique marine-mammal fauna, which includes both arctic Belugas and large oceanic Blue Whales.

The Bay of Fundy and the Gulf of Maine are marked by large tidal ranges of over 10 m. Off the coast of Nova Scotia, *The Gully* cuts in to the continental shelf.

The open northwest Atlantic Ocean

A series of deeps between the *Grand Banks of Newfoundland* and the Atlantic ridge marks the transition to the large *North American Basin*. Here, a number of plateaux are found, the *Bermuda Rise* being the largest. The southernmost part of the basin is called the *Sargasso Sea*.

Bahamas Shelf

Along Florida's Atlantic coast lies a larger shelf area which is strongly influenced by the warm Gulf Stream. The warm water permits the growth of coral reefs and has led to the creation of a series of coral islands, of which the Bahamas are the most important.

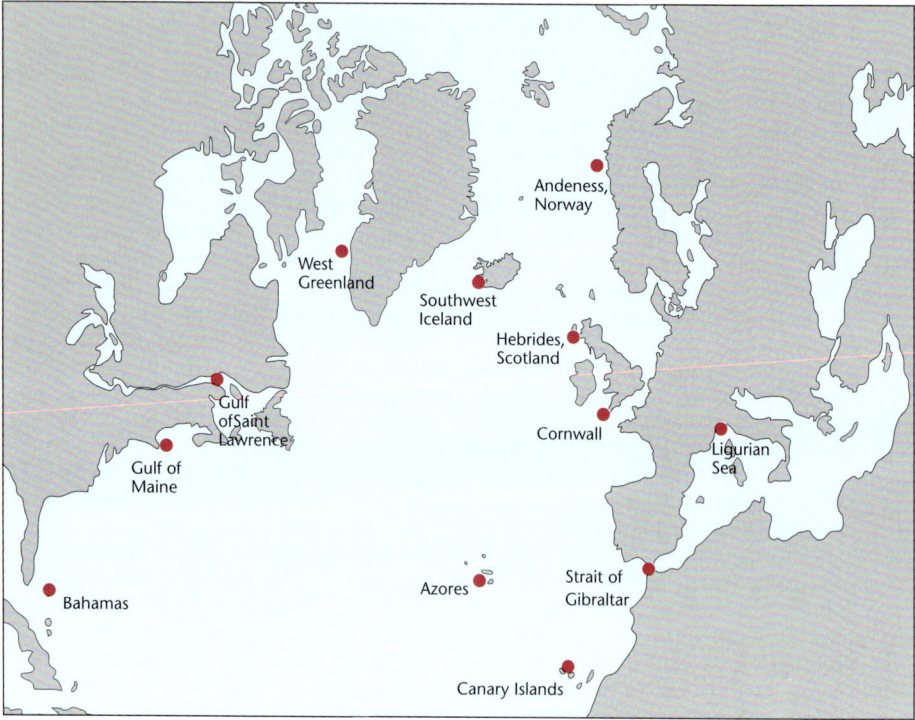

Good whale-watching localities.

Marine mammals

Marine mammals is the umbrella term for a number of mammal groups which spend all or most of their life in water—the majority of them in the sea, but a few exclusively in fresh water. The individual mammal groups are not closely related to one another. They originate from quite different land-mammal groups which have invaded the marine environment independently of each other at different points in time and have subsequently also adapted to varying degrees to life in the sea. The marine mammals include cetaceans, manatees, seals, the Polar Bear and sea otters. This book also includes coastal populations of freshwater otters in Europe and America.

The family tree of mammals and the occurrence of marine mammals. The cetaceans are linked to the even-toed ungulates, the seals to the terrestrial carnivores, and the manatees most closely to the elephants. The Polar Bear, which is the most recent offshoot, may be regarded as a Brown Bear that has changed colour and has discovered how to live in an ice-filled sea. It is 'still' a terrestrial carnivore.

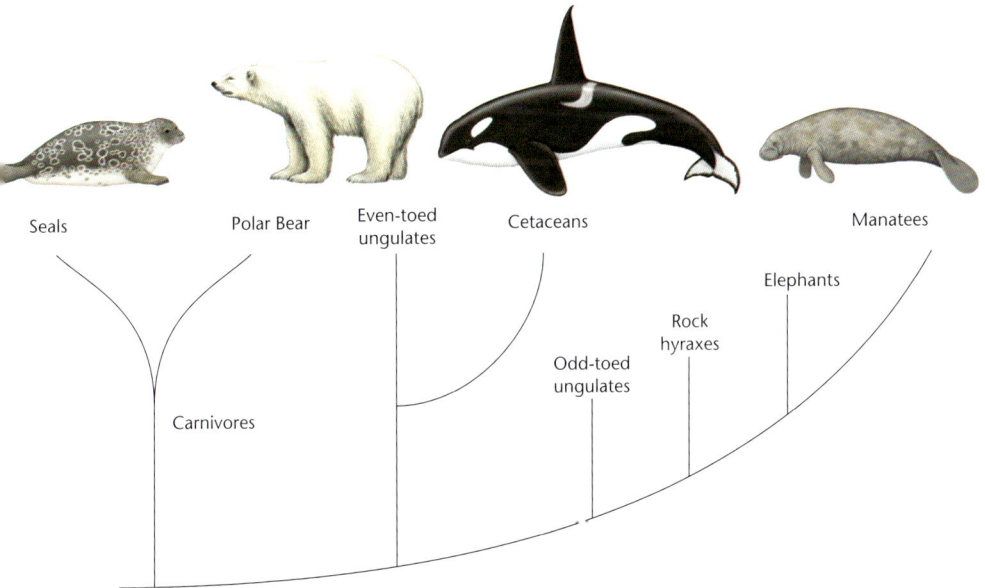

Seals   Polar Bear   Even-toed ungulates   Cetaceans   Manatees

Elephants

Rock hyraxes

Odd-toed ungulates

Carnivores

WHALES, DOLPHINS AND PORPOISES

The whales, dolphins and porpoises, or the mammalian order *Cetacea*, are adapted in the fullest possible way to life in water; cetaceans are totally independent of dry land. Births, food intake, mating and everything else take place in the water. Even though the cetaceans are called sea mammals, many of them occur more or less frequently in fresh water; indeed, four species are found exclusively in freshwater habitats.

Diversity and classification

Analysis of the relationships of organisms is known as *taxonomy*. Within this branch of research the individual species and their mutual relationships are described and examined. This could, for instance, involve clarifying whether dolphins and porpoises are closely related to each other, or if toothed and baleen whales had a common ancestor.

There are about 80 cetacean species, subdivided into toothed whales (c. 70 species) and baleen whales (c. 10 species). Some 40 of these, thus half of them, are represented in the North Atlantic. There are several reasons for resorting to approximate numbers. This is due primarily to the fact that new species are constantly being discovered in conjunction with better coverage of poorly studied sea areas and more intensive research on known species. Some species are new to science, while others emerge through the splitting of known 'species' which have been shown to consist of several latent species. The 'Common Dolphin', for example, has been shown to consist in fact of two species. Secondly, the uncertainty is due to the fact that different researchers regard the same populations as

consisting of one, two or three species (*see* table). This book follows the traditional cetacean taxonomy, while recognising that ongoing revision of the individual relationships may lead to changes in the scientific names, as well as the shifting of species between families.

The cetaceans are subdivided according to the *classification* rules applicable to the animal kingdom. This may be compared with a sorting box, where animals are sorted according to degree of similarity and relatedness into various zoological 'drawers' and 'subdrawers', or *categories*.

The cetaceans make up an *order* of the mammals, usually with two suborders recognised: *toothed whales* (Odontoceti) and *baleen whales* (Mysticeti). Some scientists, however, have developed a system with as many as five superfamilies instead, where several of the toothed whales are more closely related to the baleen whales than to the other toothed whales. In this book, the system of two suborders is followed.

Regardless of the main classification methods, related species are united in *families* consisting of *genera*. Some genera, such as the narwhal genus *Monodon*, contain only a single species, whereas, at the opposite extreme, the beaked-whale genus *Mesoplodon* consists of as many as 14 species.

The alternative classifications of right whales.

Geographical region	Presence of: one species	two species	three species
North Atlantic	Eubalaena glacialis	Eubalaena glacialis	Eubalaena glacialis
North Pacific	Eubalaena glacialis	Eubalaena glacialis	Eubalaena japonica
Southern Hemisphere	Eubalaena glacialis	Eubalaena australis	Eubalaena australis

Classification of 'Common Dolphin'. The species' scientific name consists of two Latin words: a genus name written with initial capital and a species name written with lower-case initial. The name of the person who first described the species is given after the scientific name, together with the year of description.

Category	Vertebrata	Vertebrates
Class	Mammalia	Mammals
Order	Cetacea	Whales, dolphins and porpoises
Suborder	Odontoceti	Toothed whales
Family	Delphinidae	Dolphins
Genus	Delphinus	The dolphin genus
Species	delphis	Common Dolphin *Delphinus delphis* L (= Linnaeus) 1758

Sperm whales
(3 species)

Right whales
(2–3 species)

Rorquals
(6–8 species)

Beaked whales
(20–22 species)

White whales
(2 species)

Porpoises
(6 species)

Dolphins
(33–35 species)

Pygmy whales
(1 species)

Grey whales
(1 species)

River dolphins

Baleen whales

Toothed whales

The cetaceans' family tree: only those families with members in the North Atlantic are illustrated.

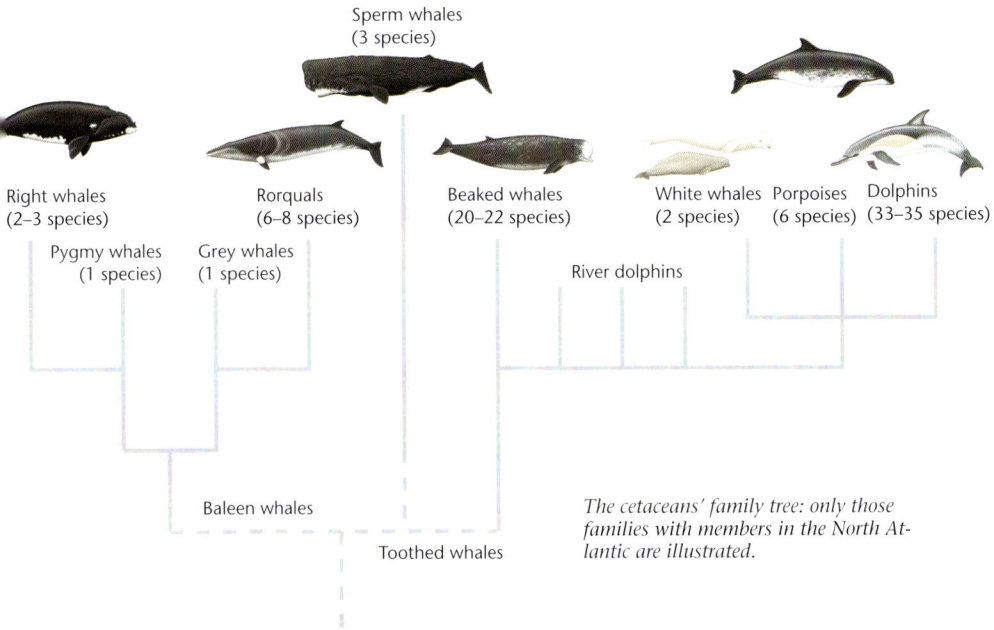

Origins and adaptations

The cetaceans evolved from primitive hoofed animals, the so-called *mesonychids*, and among the present-day mammals are closely related to the even-toed ungulates. The most recent analyses indicate that the hippopotamuses are probably the cetaceans' closest living relatives.

There has been some debate over the interrelationships of the toothed and the baleen whales. It was earlier believed that they originated from two separate groups of primitive ungulates and were not, therefore, closely related. Recent research has shown, however, that all cetaceans are more closely related to one another than they are to other animal groups.

Although the cetaceans are true mammals, they are immensely transformed mammals, being closer in appearance to a fish than to a cow.

One of the cetaceans' most striking adaptations to life in the water is that they have evolved a streamlined body. In addition, many of the external appendages either have disappeared or are hidden inside the body: external ears are totally absent; the male's scrotum is inside the body cavity, and the penis is contained within a genital slit; and the female's nipples are concealed in two slits.

Furthermore, the forelimbs are transformed into paddle-shaped flippers, while the hindlimbs have disappeared altogether (except for some vestigial bones inside the body).

Finally, the cetaceans have lost their entire pelage apart from a few sensory hairs on the upper lip.

The cetaceans' blowhole (nostrils in land mammals) is displaced to the top of the head (except on the Sperm Whale, where it is situated near the front on the left-hand side). The baleen whales have a double blowhole, and toothed whales a single one. The shifting of the nasal cavity from a normal snout position in the earliest cetaceans to a position on the top of the head is reflected in the structure of the cranium.

Physiology

The body temperature of cetaceans is the same as that of humans, about 37°C. Cetaceans often live in much colder surroundings; indeed, the difference between body temperature and the ambient temperature can be as much as 40°C in arctic seas. A layer of blubber beneath the skin insulates against heat loss, but at the same time ensures that the animals will not suffer from overheating in warmer waters. The thickest layer of blubber (up to 50 cm) is found in the Bowhead Whale, and the thinnest (a few millimetres) in some tropical dolphins.

The forelimbs lack the thick layer of blubber, and the temperature of the flippers can be considerably lower than that of the body. The body core is kept warm by means of what is known as the counterflow principle: the heat in the blood that flows out to the flippers is transferred to the cold blood flowing in from the flippers. The blood is thus warmed up before it flows back to the body (while correspondingly the outgoing blood is cooled, such that it does not lose too much heat to the surroundings).

The testicles are situated in the body cavity, the temperature of which is too high for normal sperm production. This problem is solved by a 'cooling system' very like the one mentioned above: the blood, which could overheat the testicles, is first cooled at the body's surface. Embryonic development is also heat-sensitive and could easily be inhibited by the build-up of heat which occurs in the well-insulated body, but here, too, the cooling system comes into play.

Cetaceans do not get decompression sickness (diver's disease, 'the bends'). Had they been designed like humans, their blood would dissolve large amounts of nitrogen when they dive to great depths (extreme pressure). When rising back up, they could therefore run the risk of the nitrogen creating bubbles in the blood. But

this does not happen, for cetaceans dive with the lungs empty and collapsed. The oxygen that they need when deep in the water is taken from stores in muscles and other organs, which are filled by powerful inhalation before diving.

The air and food passages are totally separate. Cetaceans can take in air, eat and swallow food simultaneously. The gullet's lowermost section is transformed into a powerful muscular grinder, and food remains are often found in this first stomach. The kidneys resemble a bunch of grapes and are efficient salt-removal organs.

Cetaceans breathe voluntarily, i.e. they have to be conscient, so they cannot sleep for lengthy periods. In most species, however, 'rest periods' occur in which the animals doze while floating at the surface (known as 'logging'). It has been demonstrated that cetaceans are able to 'sleep' with one half of the brain at a time. This guarantees the required rest and at the same time the necessary vigilance.

Living areas and migrations

The ocean is a large, continuous, but not uniform water mass. There are great differences in water depth, pressure, temperature and salinity (*see* section on North Atlantic, page 14), and the individual cetacean species live in places which are as varied as the woodlands and savannas of dry land.

These habitats can, as already mentioned, be described according to physical conditions such as sea depth, salinity, current patterns and temperature—conditions which are especially significant for cetaceans through their influence on the occurrence of prey. An insight into the nature of habitats can be gained by looking at a nautical chart. Waters cannot be judged solely by distance from land, because coastal regions can contain both large shallow-water shelf areas a great distance from oceanic conditions and deep

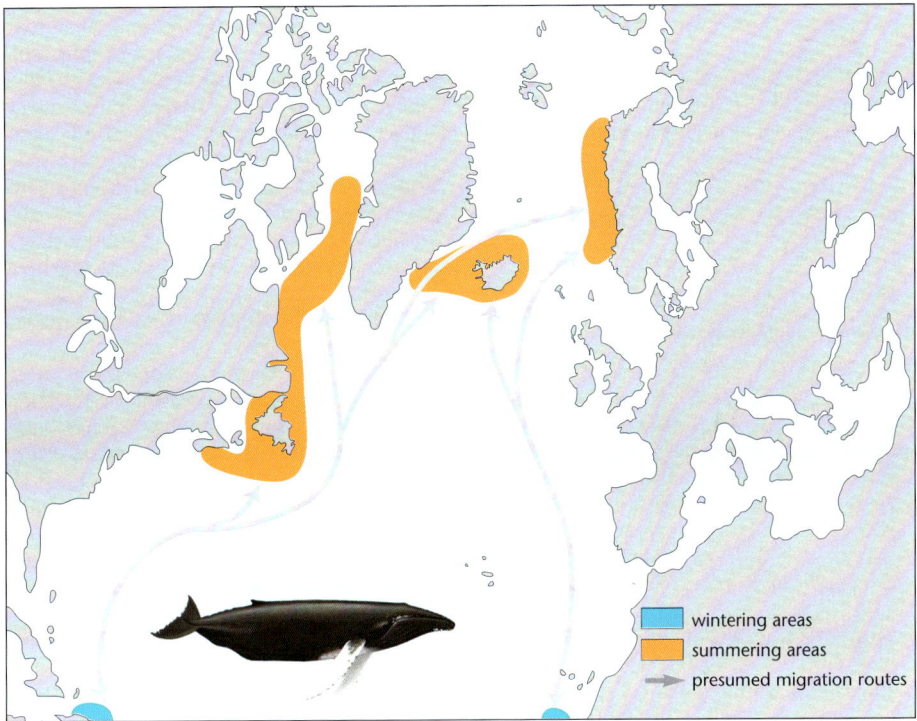

The Humpback Whale's migrations in the North Atlantic.

channels which transport oceanic water very close in to the coast.

The individual species of cetacean can be linked to different habitats or living areas. There are both 'specialists', which exploit a very restricted habitat type, and more broad-spectrum 'generalists', which are able to live in many different places. Specialised species are e.g. cetaceans which occur exclusively inshore and at times wander up rivers. Likewise, the so-called shelf species which occasionally appear along coasts are specialists, as also are the deep-sea species proper. Generalists can be found both in the open ocean over deep water and near the coast and its shallow waters. The group to which each species belongs is evident from the species descriptions which follow.

As indicated, cetaceans occur primarily where food exists for them: in other words, the occurrence of prey largely coincides with that of the cetaceans themselves.

Moreover, individuals of both sexes are found where the conditions required for mating, and often also for calving, can be met. In order to accommodate these requirements, some cetacean species undertake very well-defined migrations between foraging areas and breeding areas, while others travel constantly, like nomads, and reproduce 'en route'.

Feeding

Cetaceans are carnivores. Their food varies in size from minute crustaceans through cephalopods, fish and birds to even large baleen whales.

The baleen whales in the North Atlantic are particular in that many of them feed to a large extent on shoal fish and cephalopods rather than, as is generally the case for their relatives in the Southern Hemisphere, on crustaceans alone. There are both specialists and generalists with regard to food: some species exploit a broad spectrum of animal species, while others select a narrow range of prey.

Diving behaviour

The cetaceans spend less than a tenth of their time in the surface waters; the rest of the time they are at depth. As mentioned under 'Physiology', they do not sleep in the true sense, but for almost all species a *logging behaviour* has been described in which the animals allow themselves to drift passively at the surface while dozing.

Diving behaviour can be divided into shallow dives with comparatively frequent surfacing and deep dives with few emergences. For most species the maximum diving time or the average diving time is known, whereas the maximum diving depth for many species can only be estimated. It is also possible to examine fish and cephalopods that are found in the stomach of cetaceans. Since we often know the depths at which these prey animals occur, this provides an indirect measure of the cetacean's diving capability.

At the surface, the cetaceans draw in air a number of times and then dive again. The types of behaviour which can be recorded during the period at the surface, with breathing followed by diving and then the next surfacing event, is called the *diving sequence* and is often very characteristic for the individual species. Cetaceans swim in different ways at the surface, with some 'deeper' than others, i.e. they show less or more of the back. Some bend the back more than others when submerging, and some species raise the tail flukes when commencing a deep dive.

Besides the diving sequence itself, there are also more spectacular patterns of behaviour at the water's surface and in the air, sometimes called *aerial behaviour*. Most species, even the big whales, can leap right out of the water, but this aerial behaviour is best known among small species in the form of the *dolphin leap*, with *spinning*, in which the dolphin spins around its own axis, as the most elaborate variant. Many dolphin species also ride the bow waves of ships or ride in the wake.

In large species, the aerial behaviour is known as *breaching*. There are several variants of breaching, with a smaller or larger part of the body being hurled out of the water, and with the whale landing either on its stomach or on its back. A more sedate behaviour is *spyhopping*, in which the whale sticks only its head out of the water in order to look around above the surface.

Whales also use their flippers and tail flukes to strike the surface of the water. This is done probably both to warn of the presence of food- and mate-competitors and to deter potential enemies.

Appearance of tail flukes of Blue Whale, Bowhead Whale, Northern Right Whale, Sperm Whale and Humpback Whale when diving.

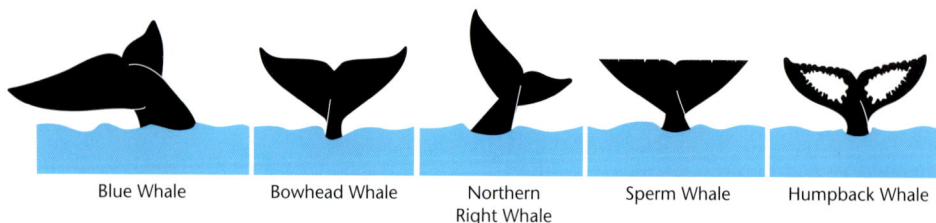

| Blue Whale | Bowhead Whale | Northern Right Whale | Sperm Whale | Humpback Whale |

Schools and pods

When several animals swim together, this is referred to as a school or pod, depending on the species. Pod structure refers to both the pod's size and composition and the duration of the association, as well as the individuals' relationship to one another.

Pods can be more or less stable units. They can involve a fixed 'core' of animals surrounded by a changing group of loosely associated animals. Or they can involve fluid social groups: i.e. animals are found together at certain times of the year or in connection with particular events such as mating, calving etc, but otherwise live solitarily.

Several pods may at times form joint groups, and several of these may combine to form *communities*. The term aggregation is used for animals which gather together for a brief period at the same place, e.g. to forage, after which they swim their own ways.

Cetacean pods can consist of both young and adult animals, and of both sexes. They can also be comprised of a single sex or a single age, e.g. only males or only juveniles, as the sexes may spend long periods apart from each other. The following examples illustrate the situation for some individual species.

In the Humpback Whale (*Megaptera novaeangliae*), the basic unit is the mother-and-calf pair. Several of these can occur in the same place. The males are predominantly solitary, but assemble in the mating areas in order to compete for the females' attention. Male Humpbacks have developed a special song (the females do not sing) which will attract the females and deter rival suitors. The males' song changes very slightly from year to year, and there are also minor geographical differences between the east and the west sides of the North Atlantic.

In the Bottlenosed Dolphin (*Tursiops truncatus*) and its close relatives, the pods consist of related females of several generations. Young males form bachelor pods, and older males may form alliances in connection with matings.

In the pilot whales (*Globicephala*), the full-grown males swim in pods together with the females, but matings take place outside the pod. The most fascinating pod structure is found in the Killer Whale (*Orcinus orca*), where the pod's members communicate by means of a special pod-specific dialect. The pods consist of related females of several generations and their offspring of both sexes. Killer Whales have evolved several sophisticated hunting methods which are based on a high degree of co-ordination and, therefore, of communication.

Whaling in the North Atlantic

There is a long tradition of whaling in the North Atlantic. Good evidence exists that cetaceans were exploited in early times on both the American and the European coasts. The activity ranged (and still ranges) from commercial exploitation to one with the purpose purely of local subsistence. A full account of all aspects is beyond the scope of this book, but the reader is referred to the bibliography on page 184.

The first commercial hunting of whales in the North Atlantic began in coastal waters off the Basque country—probably in the 1300s—and was directed at the Northern Right Whale. Coastal hunting gradually developed into hunting out at sea, and eventually was also extended to other slow-moving (and therefore named 'right' whale) species such as Bowhead Whale and, to a lesser extent, Humpbacks and the now exterminated population of North Atlantic Grey Whales. By around 1860, the populations of 'right whales' were from a commercial point of view extinct.

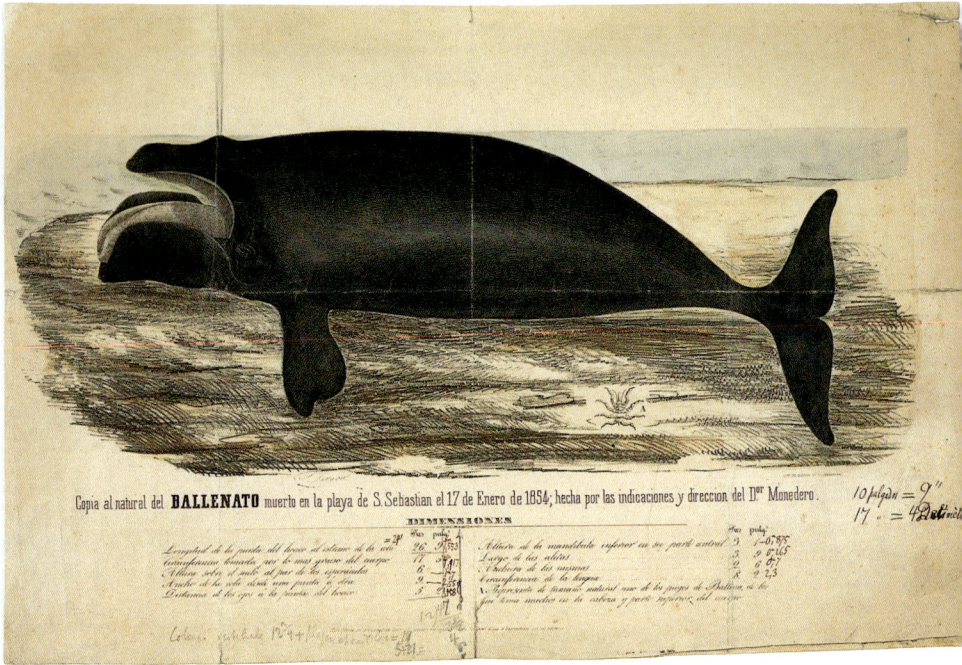

Old record of Northern Right Whale stranded at San Sebastian in 1854.

The rorquals were, until the 1860s, the 'wrong' whales, because they were too quick. And even if hunters did succeed in killing them, they did not remain floating at the surface like the right whales. This picture changed, however, with the invention of the steam ship combined with the development of the harpoon cannon. From 1868, the second big whaling boom in the North Atlantic began, initially aimed at large species such as Blue Whale and Fin Whale, but gradually also Sei and Humpback Whales.

Today, the commercial exploitation of rorquals survives as a limited hunting of Minke Whales in the Norwegian Sea.

Apart from right whales and rorquals, Sperm Whales have also been hunted periodically in the North Atlantic. The classic hunting of Sperm Whales had its heyday in the 1700s and was concentrated around

Nantucket, in New England. An offshoot of this whaling continued at the Azores right up to the early 1980s. In the 1900s, commercial hunting of Sperm Whales took place off the Spanish and Icelandic coasts.

Several smaller toothed whales have been the object of whaling. Porpoises have been caught in Denmark, along the Flemish coast and in the Black Sea, and are regularly caught locally in west Greenland. Pilot whales were killed in the Shetland Isles and Newfoundland and are still caught in the Faeroes, and Northern Bottlenose Whales and Killer Whales have likewise been hunted. In the Black Sea, hunting for Common Dolphin and Bottlenosed Dolphin still continues. In the Arctic, Beluga Whales and Narwhals are caught for local use.

Accidental by-catches

Modern fisheries represent a new threat to cetaceans in the form of accidental by-catches, chiefly of small cetaceans, in fishing gear. It is primarily the bottom nets and drift nets that are the culprits. Danish fisheries in the North Sea alone catch between 5,000 and 7,000 Harbour Porpoises annually in their gill nets.

Cetacean research

Research on cetaceans seeks, among other things, to find answers to the following questions. How can we distinguish species and populations from one another? How large are the populations? Where are the animals found in temporal and spatial

Harbour Porpoise in net.

terms? Do they migrate? What are the duration and depth of their dives? When and where do they manage to reproduce? What do they eat?

The answers to these questions rely on observations of living animals and examination of dead ones.

In very recent years, tags for satellite-tracking have been attached to individuals of a number of species, providing a wealth of new information on the animals' movements and diving behaviour.

Killing of pilot whales in the Faeroes.

Management and conservation

The earliest commercial whaling took no account of whether the cetacean populations could withstand exploitation. Had man been able to exterminate them completely, he would doubtless have done so. It gradually became clear, however, that it was necessary to limit or regulate hunting, and in 1946 the International Whaling Commission (IWC) was set up in order to regulate the hunting of large whales. The Commission has, over the 50-plus years of its history, evolved into a management body, with cetacean populations being managed on the basis of sound biological knowledge relating to their numbers, size and reproductive capacities. Recent methods of estimating populations have shown that, today, only Minke Whale populations are capable of withstanding exploitation.

The smaller cetaceans do not formally come within the IWC's sphere of authority. To compensate for this, they have been included in the Bonn Convention (CMS), which is concerned with migratory species that cross national borders. Under this convention, two international agreements have been made in recent years regarding protection of cetaceans in the North Sea and Baltic Sea (ASCOBANS) and also in the Mediterranean Sea and adjoining waters (ACCOBAMS). Under these agreements, the signatory countries commit themselves

Sperm Whale diving just off a ship carrying tourists.

to protecting their cetaceans by initiating a large number of research, management and promotional activities. The agreements have, among other things, contributed to the creation of the first European reserves for Harbour Porpoise, at Sylt in Germany, and for a number of dolphins and large whales, in the Ligurian Sea between Corsica and the French and Italian Riviera.

A special organisation for regulating whaling and sealing, the North Atlantic Marine Mammal Commission (NAMM-CO), was established in 1992.

Whale tourism and whale-watching

At various places in the North Atlantic organised whale tourism now takes place. This occurs in the Gulf of Saint Lawrence and the mouth of the Saint Lawrence River, on the coasts of Newfoundland and New England, at the Florida coast and the Bahamas, in Norway, Greenland and Iceland, in the Faeroes, the Canary Islands and the Azores, and in the Mediterranean Sea.

Besides the organised tours, it is possible to go whale-watching on one's own, as cetaceans can, in principle, be seen anywhere in the North Atlantic, from land and at sea. More detailed information on the possibilities for encountering the individual cetacean species are given in the species accounts. A list of further literature is given at the end of the book.

Cetacean sightings and field identification

Broadly speaking, there are two ways of encountering cetaceans: either as living animals in the water or as dead ones on the shore or floating at the water's surface.

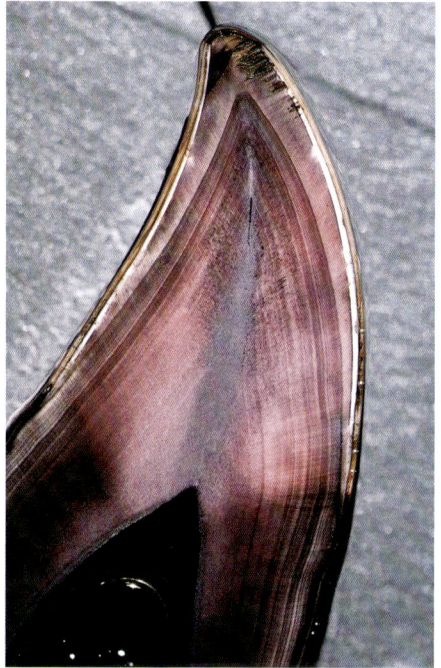

Cross-section of tooth of pilot whale showing growth layers. Characteristics of this kind can be examined only on dead animals and can be used to determine age.

Strandings have the advantage that there is time to examine the animal and to look for even very small or otherwise concealed features. This applies e.g. to the beaked whales, which are distinguished from one another mainly by the teeth; in females and juveniles of these species the teeth are often not visible at all outside the jaw, and in these cases a reliable identification is possible only if the animal can be examined. But stranded animals also present problems, as external characters disappear with increasing decomposition. In such cases, one may be forced to turn to the skull or other bones for a positive identification.

Observations of live cetaceans in the water have the drawback that the animals are seen only briefly, perhaps even in poor weather conditions or from the deck of a rolling ship. The longer the animal is observed and the closer the range, the greater is the chance of determining the species to which it belongs.

Identification of a species is achieved by the positive or the negative method. With positive identification, the species' identity is established by using characteristics which apply unambiguously to the species in question. With negative identification, the elimination method is used: i.e. a particular species is identified as such because it cannot be any of the others.

Positive identification of species is a prerequisite for any scientific investigation. Studies of e.g. ecology and behaviour, for instance, are not of much value when there is doubt as to the species on which the studies are being carried out.

Field characters

When in the field, the following key features should be noted:

– How big is the animal?
– What colour(s) is it?
– Is there a dorsal fin? Where is this positioned on the back?
– What does the dorsal fin look like?
– Does the cetacean show the tail flukes when diving?
– Does it have a beak?

It is important not to over-interpret but to identify species only from the facts.

Besides these characters, which can be difficult to see because most of the animal is underwater for most of the time, the 'aerial behaviour' or diving pattern of different species can provide important factual evidence for a specific identification. In the case of the larger species, the appearance of the blow can also play a part.

In some species there is a significant difference between males and females. In other species the males have what are

Blow of some large whales.

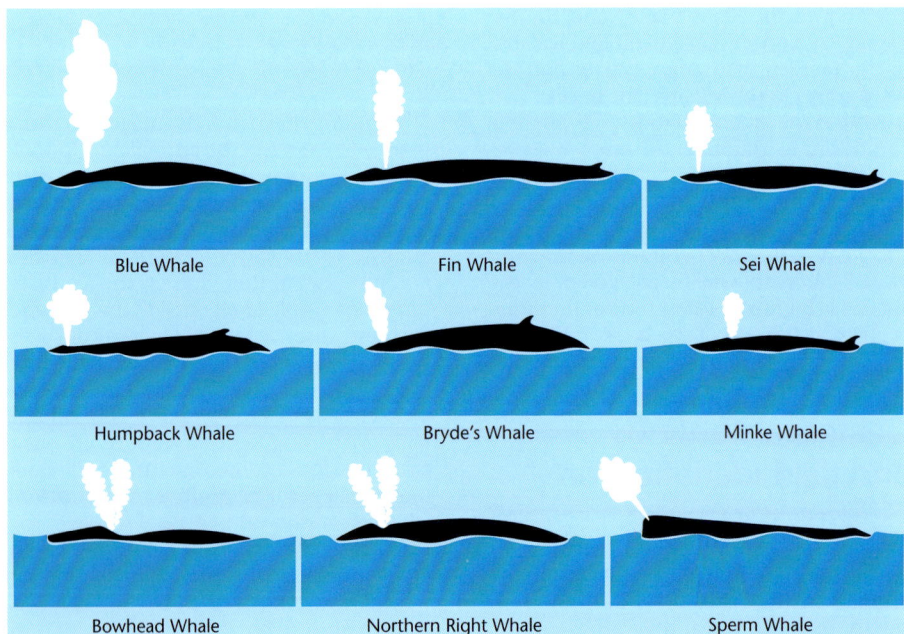

Blue Whale　　　Fin Whale　　　Sei Whale

Humpback Whale　　　Bryde's Whale　　　Minke Whale

Bowhead Whale　　　Northern Right Whale　　　Sperm Whale

Morphological terms of cetaceans.

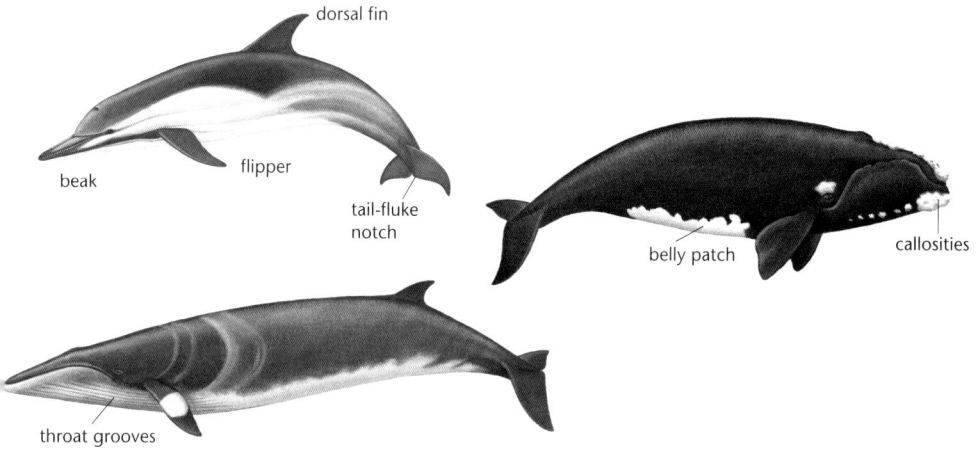

dorsal fin

beak

flipper

tail-fluke notch

belly patch

callosities

throat grooves

Sexing of whales.

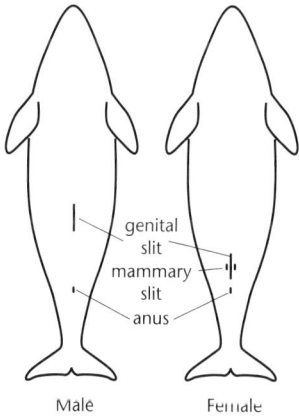

genital slit

mammary slit

anus

Male

Female

Variation in shape and positioning of dorsal fin.

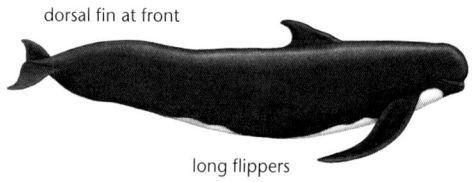

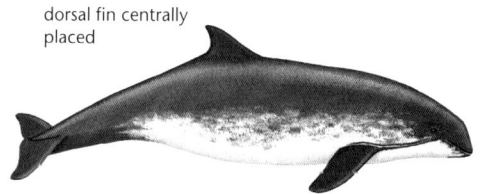

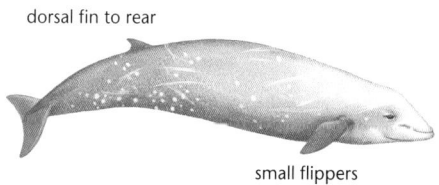

dorsal fin at front

long flippers

dorsal fin centrally placed

dorsal fin to rear

small flippers

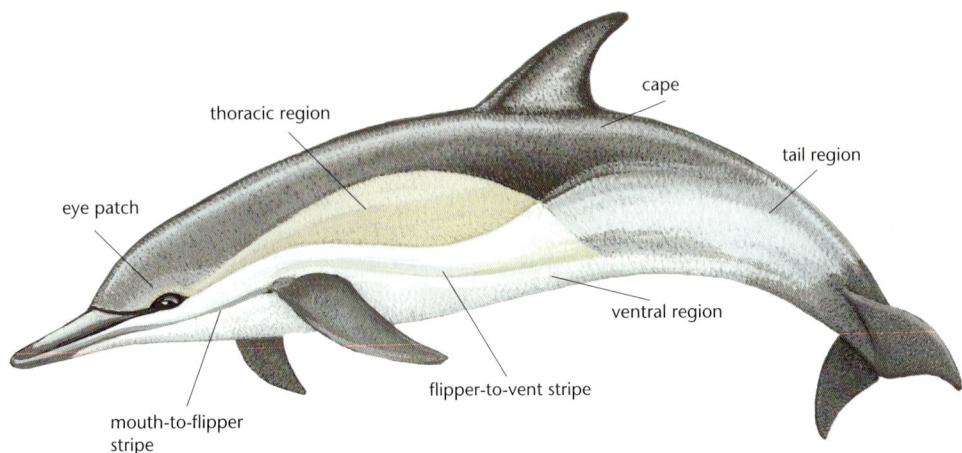

Colour markings of cetaceans.

termed secondary sexual characteristics (e.g. tusks).

A number of technical terms are used in describing a cetacean's appearance. The nasal openings are called the *blowhole* and the snout the *beak*. The swollen part of the head behind the beak is called the *melon*. The paddle-shaped pectoral fins are referred to as *flippers*. On the belly, the genital openings are in a fold known as the *genital slit*. On females, the nipples are situated in *mammary slits*.

The colour patterns are important characters. Most cetaceans are coloured according to the countershade principle, with a *dark back* and a *pale belly* (making them difficult to see both from above and from below). On the sides, the transition between the back and the belly colours is either sharply or diffusely defined; sometimes there is a *transitional panel* of intermediate

tone. If the back has an area different in colour from the rest, this is referred to as a *cape*.

The three basic areas—back, belly and sides—may be broken up either by a pattern of spots or by the occurrence of stripes or patches of various shapes. On some species this involves only a difference in shade, whereas others exhibit a sharply defined contrast. Young animals of some species have a colour pattern that differs from that of the adults.

The colour pattern of some dolphins is more complex. The panels of colour can be in different positions and vary in intensity, such that a darker patch may completely cover a lighter one, or an underlying patch may show through a 'transparent' overlying one. Basically, however, the pattern of a dark back, paler sides and a white belly still applies.

Toothed whales Odontoceti

The suborder of toothed whales traditionally comprises eight families, five of which occur in the North Atlantic. These are the dolphin family (Delphinidae), the porpoise family (Phocoenidae), the Beluga and Narwhal family (Monodontidae), the beaked whale family (Ziphiidae) and the sperm whale family (Physeteridae). The suborder thus comprises small, medium-sized and large cetacean species.

Common to all the toothed whales is, of course, the presence of teeth, though these vary greatly in shape, function and number. In addition, toothed whales have only a single blowhole, which is situated on the top of the head except in the case of the Sperm Whale (*Physeter macrocephalus*), where it is at the front on the left-hand side. The exhaled air, the blow, is generally visible only in the larger species.

Dolphins

The dolphin family (Delphinidae), with 33 species worldwide, is the largest of the order Cetacea. The family comprises both large and small species, some with and some without a beak, and species with few

Long and short dolphin beaks.

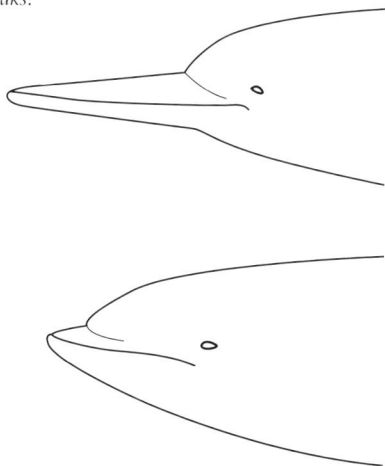

teeth and others with many. It is traditionally divided into several subfamilies. In this book the traditional classification is followed, although this does not necessarily reflect the true relationships among these species. The relationships among the individual dolphin species and subfamilies are rather complicated, and only in recent years has some light been shed on these. For a more thorough treatment of relationships, refer to the bibliography at the end of the book.

Three subfamilies, the true dolphins, the rough-toothed dolphins, and the pilot and killer whales, occur in the North Atlantic, where a total of 20 dolphin species has been recorded.

True dolphins

The true dolphins (Delphininae) are also known locally as 'jumpers', from their characteristic habit of leaping. The 11 species united in this group all have a distinctive beak, although this varies considerably in length. It is short (up to 6 cm) on species such as the White-beaked (*Lagenorhynchus albirostris*), White-sided (*Lagenorhynchus*

Short-beaked Common Dolphin, page 48

Pantropical Spotted Dolphin, page 58

Long-beaked Common Dolphin, page 51

Clymene Dolphin, page 60

Bottlenosed Dolphin, page 53

Spinner Dolphin, page 62

Atlantic Spotted Dolphin, page 55

Striped Dolphin, page 65

Comparative plate of dolphins.

acutus) and Fraser's Dolphins (*Lagenodelphis hosei*), of medium length on the Bottlenosed Dolphin (*Tursiops truncatus*), the spotted dolphins (*Stenella attenuata* and *Stenella frontalis*), the Short-beaked Common Dolphin (*Delphinus delphis*) and the Clymene Dolphin (*Stenella clymene*), and longest (up to 30 cm) on the Long-beaked Common Dolphin (*Delphinus capensis*) and the Spinner Dolphin (*Stenella longirostris*).

Two species, the White-beaked and White-sided Dolphins, are found only in the North Atlantic, and a further two, the Atlantic Spotted and the Clymene Dolphins, are found only in the subtropical and tropical Atlantic seas. The other species have wider distributions in tropical to warm-temperate waters.

The North Atlantic is therefore unique for the occurrence of two species of spotted dolphin and two species of spinner dolphin (the Spinner and the Clymene Dolphins).

The distribution of some species in the North Atlantic is only approximately known. As these species could turn up 'unexpectedly' in a region where one is watching, identifications made on the basis of 'geographical elimination' (*see* page 12) are too uncertain. In addition, variation within the individual species is still insufficiently understood or is known only for other, better-studied regions.

Some species have both coastal populations and populations that prefer the open ocean. Such populations are known as *ecotypes* and they can, as a rule, be distinguished by structural characteristics. The splitting of species into ecologically defined populations, the ecotypes, may be regarded as the first stage in the creation of new species. Whereas common dolphins (genus *Delphinus*) are, as mentioned below, already split into two species, the speciation process is perhaps not yet so advanced in the case of the Bottlenosed Dolphin (*Tursiops truncatus*), which similarly occurs in both coastal and oceanic forms.

The 'Common Dolphin' was for many years considered a single, highly variable species (*Delphinus delphis*) with many geographical forms and races. In 1994, however, a more detailed investigation revealed that two species were involved: the Short-beaked Common Dolphin, which retained the scientific name *Delphinus delphis*, and the Long-beaked Common Dolphin, which was given the name *Delphinus capensis*.

Both species occur in the North Atlantic, but, as old reports do not distinguish between them, the two species' distributions are only now being mapped in detail. The Short-beaked species is so far the only one that has been recorded in the Mediterranean Sea, the Black Sea and along north European coasts, whereas both species are found on the West African coast and the east coast of the USA. The Long-beaked Common Dolphin is, moreover, found far out at sea, including around the Azores. The two species' habitats probably overlap each other to some extent, but the Short-beaked species seems to prefer shelf and coastal regions, while the Long-beaked seems to have both a more oceanic and a more southerly distribution.

The comparative plates (*see* page 42) are intended as an aid to identification. It is, however, important to state that one *cannot* expect to identify to species level every single dolphin that one sees. Guessing should be avoided and you should stick strictly to what you have observed, even if this means that you have seen 'only' an unspecified long-beaked dolphin (*Stenella longirostris/Delphinus capensis*) or a White-beaked/White-sided/Bottlenosed Dolphin.

White-beaked Dolphin.

White-beaked Dolphin
Lagenorhynchus albirostris

FRENCH Dauphin à bec blanc
GERMAN Weißschnauzen-Delphin
SPANISH Delfín de hocico blanco

Description: Robust short-beaked dolphin with rather tall centrally placed dorsal fin.
Size: Full-grown adults are 2.4–3 m long and weigh 180–350 kg; calves are 1.2 m long and weigh c. 40 kg.
Teeth: There are 25–28 conical teeth in each half of both jaws.

Coloration: Characteristic white beak. Black back with white saddle behind dorsal fin. White bands of varying intensity on flanks—either as proper patches or as greyish tones. Belly and beak are usually white.
Variation: Males are slightly bigger than females. On some individuals, the white beak can be dirty grey, or even so dark that the animal appears to lack the white beak altogether.

Distribution: Occurs in temperate and subarctic coastal and shelf waters of the North Atlantic. In the northeast found as far north as the White Sea. It is common on the Norwegian west coast. In the North Sea, it is seen most frequently in summer. Fairly frequent in the Skagerrak and the Kattegat, and occasionally moves right into the Baltic. Has been recorded in recent years on the French and Spanish coasts.

Southernmost coasts of Portugal probably represent the southern limit of distribution in the eastern North Atlantic. In the western part, it is most common around Labrador and southwest Greenland. The species nomally occurs as far south as Cape Cod Bay.

White-beaked Dolphins.

Behaviour: Pods of up to 30 animals are not uncommon, but sometimes several hundred or even several thousand animals are seen together. Details of the composition (structure) of groups are not known.

Can be inquisitive and approach smaller vessels. Very acrobatic behaviour with typical dolphin leaps. In good observation conditions, the ripples made by the tail flukes can be seen. Occasionally rides bow-waves of ships. Diving behaviour is not known in detail.

Sometimes swims with rorquals and forms mixed schools with White-sided Dolphins.

Reproduction: Not particularly well known. Sexually mature females are c. 2.4 m long, while males become sexually mature at a length of c. 2.5 m. Insufficient is known about the age at sexual maturity.

Mating takes place in summer. The gestation period is not known, but is assumed to be 10–11 months as in some other dolphins. Calving takes place in summer. There is no information on the duration of lactation. In principle, the mature female should be able to give birth to one young every year, but there are probably intervals between pregnancies.

Food: Codfish and cephalopods.

Natural enemies: Presumably Killer Whales and, in the case of young, perhaps large shark species.

Relationship with man: There is no commercial exploitation of the species. Occasional individuals are killed in southwest Greenland. Accidental by-catches occur both in bottom gill nets and in trawls.

Similar species: The only other species of similar size and with the same small beak is the *White-sided Dolphin*, but that has neither a white saddle nor a white beak. The White-sided Dolphin has an olive-yellow band on the rear flanks.

Fraser's Dolphin has the same size and robust body shape, but the two species' ranges do not normally overlap.

Owing to its sharp black-and-white colour markings and its comparatively tall dorsal fin, the White-beaked Dolphin can sometimes be confused with the *Killer Whale*. The latter, however, is much bigger, and once the Killer Whale's dorsal fin has been seen there is rarely any doubt. The White-beaked Dolphin's aerial behaviour is rather different from the Killer Whale's, with far more frequent leaps.

White-sided Dolphin
Lagenorhynchus acutus

White-sided Dolphin.

FRENCH Dauphin à flancs blancs de l'Atlantique
GERMAN Weißseitendelphin
SPANISH Delfín de flancos blancos

Description: Slender, short-beaked dolphin with rather tall centrally placed dorsal fin.
Size: Full-grown adults are 1.9–2.9 m long and weigh 165–200 kg; calves are 1.2 m long and weigh c. 35 kg.

Teeth: There are 29–40 conical teeth in each half of both jaws.
Coloration: The back is black. The flanks have a white band, which at the rear runs into an olive-yellow stripe. The beak is black.
Variation: Males are slightly bigger than females. The olive-yellow band can be rather faded and therefore does not always appear distinctive in the field.

Distribution: Temperate and subarctic oceanic parts of North Atlantic. In the eastern North Atlantic from Svalbard and the White Sea in the north, and south along the Norwegian coast. Occurs along Scotland's north coast, the Shetland Isles, the Faeroes and Iceland. In the south, there are occasional records from the Biscay coast of Spain and France. In the western North Atlantic it is quite common along USA coasts, as the shelf edge reaches rather close to land. Known also from west Greenland and Labrador.

Behaviour: Pods normally consist of up to 50 animals, but much bigger groups of up to several thousands are observed in the middle of the North Atlantic.

Acrobatic, but is rather shy and does not approach ships. Takes breaths every 15–20 seconds, but does not always leap out of the water. Sometimes sticks snout out of the water. Details of diving behaviour are not known.

Forms groups with White-beaked Dolphins, and sometimes swims with rorquals.

Reproduction: Males become mature at a length of 2.2 m, females when 2 m long. The age at sexual maturity is not clear.

Mating occurs in late summer. Gestation lasts 11 months. Births take place from May to August, but peak in June and July. Lactation can last for as long as 2 years, but females have also been found which were pregnant at the same time as they were suckling a calf.

Food: Pelagic fish and cephalopods.

Natural enemies: Presumably Killer Whales.

Relationship with man: No commercial exploitation, but hunting occurs locally in southwest Greenland, Labrador and the Faeroes.

Similar species: The only other species of roughly similar size and with the same small beak is the *White-beaked Dolphin*, but that has a white saddle and a white beak.

Fraser's Dolphin is similar in size and has the same robust body shape, but the two species' ranges do not normally overlap.

Because of the contrasting black-and-white markings and the relatively tall dorsal fin, the White-sided Dolphin can perhaps be confused with the *Killer Whale*. The Killer Whale, however, is much bigger, and the mere sight of its dorsal fin removes all doubts. The White-sided Dolphin's aerial behaviour is also more pronounced than the Killer Whale's.

White-sided Dolphin.

Flotilla of Short-beaked Common Dolphins.

Short-beaked Common Dolphin
Delphinus delphis

FRENCH Dauphin commun à bec brève
GERMAN Kurzschnauziger Echtdelphin
SPANISH Delfín común a bec brava

Description: Slender dolphin with clearly demarcated beak.

Size: Full-grown adults are 2–2.5 m long and weigh 70–135 kg; calves are 80 cm long and weigh up to 10 kg.

Teeth: There are 40–61 teeth in each half of both jaws.

Coloration: A well-defined hourglass pattern on the flanks with yellow panel at front and grey panel at rear, a black cape and a white belly. The cape forms a characteristic point on the flank, level with the dorsal fin. The head has a black eye patch, and there is a black stripe from the corner of the mouth to the flipper. The back is black. The beak is black as a rule, but can be pale-tipped. Dorsal fin can vary from all black to almost white.

Variation: Males are a little bigger than females. Some individuals have a black-bordered grey panel in the centre of the dorsal fin. One or more yellowish stripes running from the flipper to the vent.

Behaviour: Pods of several hundred animals, sometimes thousands. The size of groups varies with time of day and time of year. Precise composition of groups is not known in detail, but is perhaps similar to that of other, better-studied species. Stray individuals or small pods may occur in the northern part of the range. There are no well-defined migrations known to take

Distribution: Tropical, subtropical and warm-temperate seas with temperatures between 10°C and 28°C. Both pelagic and coastal, sometimes occurring on the Norwegian coast and in the North Sea, and even in the Kattegat. Rather common in the western and central Mediterranean. An independent population is found in the Black Sea.

In the western North Atlantic, the species occurs along the American east coast from Cape Hatteras in the south to Newfoundland in the north. It appears more rarely farther south, e.g. on the Florida coast.

The Short-beaked Common Dolphin is predominantly a coastal species, and there are even reports of individuals which have swum several kilometres up rivers.

place. Some groups remain in the same area throughout the year, whereas others are better described as nomadic. Groups pack together at signs of danger.

These dolphins ride the bow-waves of both ships and large whales. They can dive to 280 m and remain submerged for 8 minutes. More commonly, however, they remain below the surface for between 2 seconds and 2 minutes. Dive duration also depends on whether the animal is travelling or whether it is foraging. No mass strandings are known for this species. Larger numbers of stranded or dead dolphins are the result of viral epidemics or accidental by-catches.

Mixed schools of Common Dolphins and other species occur.

Reproduction: Males become sexually mature when c. 2 m long (1.75 m in the Black Sea), females when 1.9 m long

Short-beaked Common Dolphin.

Two Short-beaked Common Dolphins.

(1.6 m in Black Sea). The age of maturity is 5–7 years for males, 8–14 years for females.

Mating can take place throughout the year, even though seasonal changes in testicular activity have been demonstrated. Gestation lasts 10–11 months. Births, too, can take place throughout the year, although, in some places, spring or summer–winter is said to be the main calving season. The calf is nursed by the mother for a rather long period, perhaps up to 1 year. Births can, in theory, take place every year, but rest years occur. Twin births are known, but are rather rare.

Food: Dolphins eat a wide range of shoaling fish and cephalopods, and the composition of the diet varies geographically and seasonally.

Natural enemies: Presumably Killer Whales and larger species of shark.

Relationship with man: Some hunting of the Black Sea population still occurs. Apart from this, there is no commercial exploitation, but accidental by-catches occur. Dolphins from several North Atlantic populations are kept in captivity.

Similar species: The Short-beaked Common Dolphin can be confused with its Long-beaked relative and with all *Stenella* species occurring in the region—Striped, Spinner and Clymene Dolphins and Pantropical and Atlantic Spotted Dolphins—and in certain situations with small Bottlenosed Dolphins.

Compared with the *Long-beaked Common Dolphin*, the beak is shorter and there are differences in the colour markings of the head and body. Compared with the *Striped Dolphin* and the two species of *spotted dolphin*, the colour pattern is different. *Spinner* and *Clymene Dolphins* differ in behaviour. Differs from small *Bottlenosed Dolphins* in colour markings, body shape and beak length. *See also* the comparative plate on page 42.

Long-beaked Common Dolphin.

Long-beaked Common Dolphin
Delphinus capensis

FRENCH Dauphin commun à bec long
GERMAN Langschnauziger Echtdelphin
SPANISH Delfín común a bec longo

Description: Slender dolphin with clearly demarcated long beak.
Size: Full-grown adults are 2.2–2.7 m long and weigh 80–135 kg; calves are 80 cm long and weigh up to 10 kg.
Teeth: 53–61 teeth in each half of both jaws.

Coloration: An obscure hourglass pattern on the flanks with yellow panel at front and grey panel at rear, a black cape and a white belly. The head has a black eye patch and a black stripe from the corner of the mouth to the flipper. The back is black, though the dorsal fin can be white at rear. Variations in colour markings are not known in detail. Normally has a broad black patch on the lower jaw.
Variation: Males slightly bigger than females.

Distribution: Tropical, subtropical and warm-temperate seas with temperatures between 10°C and 28°C. Pelagic and coastal. Occurs in southern parts of North Atlantic, where it has been recorded on the northeast coast of Brazil, the west coast of Africa and around the Azores.

Behaviour: Probably does not differ much from the Short-beaked species. Appears in groups of several hundred animals, sometimes thousands. Composition of groups is not known in detail, but is perhaps similar to that of other, better-studied species. In the northern part of the range, stray individuals or small pods can occur. No well-defined migrations are known.

Rides the bow-waves of ships and also those made by large whales.

Mixed schools of Common Dolphins and other dolphin species occur.

Reproduction: There is no detailed information on the breeding of this species in the North Atlantic. Off the Californian coast, the length at maturity is given as 2.02–2.35 m for males and 1.93–2.24 m for females, but it is not known whether these figures are also valid for the North Atlantic. Timing of mating and calving are not known for the North Atlantic but, if they follow the general pattern, they are year-round events. Gestation presumably lasts 10–11 months.

Food: Not known, but probably consists of oceanic fish and cephalopods.

Natural enemies: Presumably Killer Whales and large sharks.

Relationship with man: There is no commercial exploitation of this species, but accidental by-catches in fishing gear can be expected.

Similar species: The Long-beaked Common Dolphin can be confused with its Short-beaked relative and with all *Stenella* species occurring in the region—Striped, Spinner and Clymene Dolphins and Pantropical and Atlantic Spotted Dolphins—and in certain situations with small Bottlenosed Dolphins.

Compared with the *Short-beaked Common Dolphin*, the beak is longer and there are differences in the colour markings of the head and body. Compared with the *Striped Dolphin* and the two *spotted dolphins*, the colour pattern is different. *Spinner* and *Clymene Dolphins* differ in behaviour. Differs from small *Bottlenosed Dolphins* in colour markings, body shape and beak length. *See also* the comparative plate on page 42.

Long-beaked Common Dolphin.

Bottlenosed Dolphin.

Bottlenosed Dolphin
Tursiops truncatus

FRENCH Grand dauphin
GERMAN Großtümler
SPANISH Tursion, Calderon

Description: Relatively robust dolphin with medium-length beak, but slimmer tropical forms with longer beak are also known.
Size: Full-grown adults are 1.9–4 m long and weigh 150–650 kg (smaller sub-tropical/tropical forms up to 2.5 m and 250 kg); calves are 1 m long and weigh 15–30 kg.
Teeth: 19–24 medium-sized teeth in each half of both jaws.
Coloration: Dark grey back, light grey flanks and white belly; belly sometimes pink. The cape can be very distinct, so that, from above, the animal is con-spicuous with a prominent dark band. Old females can have spots on the belly. Tropical populations have a more bluish-black back, and spots on the underside.

Distribution: Occurs both inshore and at sea in North Atlantic tropical, sub-tropical and warm-temperate waters.

In the eastern North Atlantic, the northern limit of distribution is the English Channel and north Scotland. A population was formerly found in the southern North Sea. Elsewhere, strays have been recorded along the coasts of Iceland, the Faeroes and Norway and in the Baltic Sea. The Bottlenosed Dolphin is quite common in the Mediterranean Sea, with well-studied populations in e.g. the Adriatic Sea. An isolated population is found in the Black Sea. Known also from the West African coast and the Canary Islands, Madeira and the Azores.

Along the American east coast, the coastal form occurs north to Cape Hat-teras, while the oceanic form occurs even farther north, with the Gulf Stream probably the northern limit.

Bottlenosed Dolphin.

Variation: Males bigger than females. The Bottlenosed Dolphin is a geographically highly variable species. There are both coastal and oceanic forms, large and small forms, and long-beaked and short-beaked forms. Some forms are to a greater or lesser degree spotted on the belly. Unlike the Common Dolphins, the Bottlenosed Dolphin's various populations are not split into separate species.

European coasts harbour rather large animals of up to 4 m, whereas those on the American coast are much smaller, with a maximum length of 2.5 m. The Black Sea's Bottlenosed Dolphins are intermediate in size between these two extremes. Current research on animals from the Mauritanian and Senegal coasts suggests that the dolphins there also belong to the large form. It is not known whether a smaller and more spotted tropical form occurs farther south, e.g. off Ivory Coast and Ghana; if it does, such individuals may occasionally be observed at the southern end of the region covered by this book.

Behaviour: The behaviour of the Bottle-nosed Dolphin is among the best-studied of all cetacean species, because it has been kept in captivity since the late 1930s.

Where group structure is known, it involves both female pods with young of both sexes and pods containing young males. There are normally up to 10 animals in a pod in coastal forms, whereas pods of the offshore populations are slightly larger. Extensive co-operation occurs within the group.

In certain places along the coast, the animals have developed a special foraging method which has proven to be most successful in terms of numbers of fish caught: they chase the fish up on to the shore by stranding themselves, and then wriggle their way back to the water with a 'mouth full of food'. Seldom dives for longer than 4 minutes at a time. Presumed to be capable of diving to several hundred metres and remaining submerged for more than 20 minutes.

Reproduction: Males become sexually mature at an age of 8–12 years and females at 5–10 years. Length at onset of maturity varies considerably: 1.9–2.4 m for females and 2.1–2.6 m for males. Gestation lasts 12 months and is followed by a lactation period of 12–18 months.

Food: A variety of fish and cephalopods.

Natural enemies: Presumably Killer Whales and large sharks.

Relationship with man: Direct hunting has taken place only in the Black Sea. Accidental by-catches occur in many places within its range. The Bottlenosed Dolphin has been kept in captivity since the end of the 1930s, and is probably the dolphin known to most people by the name 'Flipper'.

Similar species: Its size and the well-pronounced medium-length beak mean that the Bottlenosed Dolphin is normally not confused with other dolphins at short range. At longer distances, however, confusion can arise with other *medium-beaked dolphins*.

In tropical waters it can be difficult to distinguish small Bottlenosed Dolphins from the *Atlantic Spotted Dolphin*, as this can have very poorly developed spots. The *Rough-toothed Dolphin* resembles the Bottlenosed in coloration, but is distinguished by the head shape, with a beak that is not well defined but merges gradually into the head.

Atlantic Spotted Dolphin just beneath surface.

Atlantic Spotted Dolphin
Stenella frontalis

FRENCH Dauphin tacheté de l'Atlantique
GERMAN Atlantischer Pfleckendelphin
SPANISH Delfín pintado

Description: Relatively 'compact' dolphin with medium-length but robust beak.
Size: Full-grown adults are 1.7–2.3 m long and weigh 100–143 kg; calves are 80–120 cm at birth.
Teeth: 30–42 teeth in each half of both jaws.
Coloration: Typical young animals have a black or dark cape and a greyish 'sideways V' on the flanks. Black spots gradually form on the white belly and white spots on the flanks and the back. The spots can completely cover the dark cape, the grey central band and the white belly.
Variation: Males are a little bigger than females. The animals are born without spots, which appear gradually. Populations of almost unspotted animals exist, however, which complicates the situation. There are coastal and oceanic populations of the species.

Atlantic Spotted Dolphin.

Behaviour: Probably forms complex groups in the manner of other dolphin species. Normally fewer than 50 individuals in a pod; pods appearing close inshore contain only 5–15 animals.

Very active dolphin. Is attracted to fast-moving ships and rides their bow-waves. Diving behaviour is not described in detail, but is presumably very similar to that of other small dolphins in the region.

Interactions with Bottlenosed Dolphins occur, and forms mixed schools with that species. There are even records of hybrids between the two.

Reproduction: Very poorly known. Females probably reach sexual maturity when 1.8–1.9 m long, males at 2.1–2.2 m, but there seems to be significant geographical variation among populations. The age at sexual maturity is not known. Gestation is presumed to last 10–12 months. Insufficient information is available on the duration of the lactation period.

Food: Mainly fish.

Natural enemies: Large sharks and probably Killer Whales.

Relationship with man: Accidental by-catches occur in various kinds of fishing gear, chiefly bottom nets and drift nets and also purse seine. Several individuals

Distribution: Occurs out at sea in the southern part of the North Atlantic, with the northernmost limit at 45°N. Observed at the Canary Islands, the Azores and the West African coast. Occurs on the American east coast, in summer as far north as New England. Well-studied populations are found in the Bahamas.

have been kept in captivity, but have survived for only a short time.

Similar species: Very similar to the *Bottlenosed Dolphin* owing to the relatively short and broad snout. It may be necessary to check again to see whether any Bottlenosed Dolphins are present in a school, or whether just juveniles or weakly spotted individuals are involved. In addition, Bottlenosed Dolphins can be spotted on the belly and flanks. The *Pantropical Spotted Dolphin* is also spotted, but has a very different ground coloration and is slimmer. The *Clymene Dolphin*, *Striped Dolphin* and *Short-beaked Common Dolphin*, when seen at longer ranges and against the light, can resemble this species. The *Spinner Dolphin* and *Long-beaked Common Dolphin* are distinguishable by the very long beak.

Pod of Atlantic Spotted Dolphins.

Pantropical Spotted Dolphin

Stenella attenuata

FRENCH Dauphin tacheté pantropical
GERMAN Pantropischer Tüpfeldelphin
SPANISH Delfín moteado

Description: Slim dolphin with well-demarcated beak and centrally positioned dorsal fin.
Size: Full-grown adults are 1.7–2.6 m long and weigh 90–120 kg; calves 80–90 cm, with weight at birth probably c. 10 kg.

Pantropical Spotted Dolphin.

Teeth: There are 34–47 teeth in each half of both jaws.
Coloration: The spots become larger and more numerous with age and finally merge together, so that the spotted areas appear uniformly grey. Tip of beak often white.
Variation: Full-grown males normally reach up to 2.07 m in length, and females 1.95 m. Males are a little bigger than females, which is due to a greater growth rate after the onset of sexual maturity.

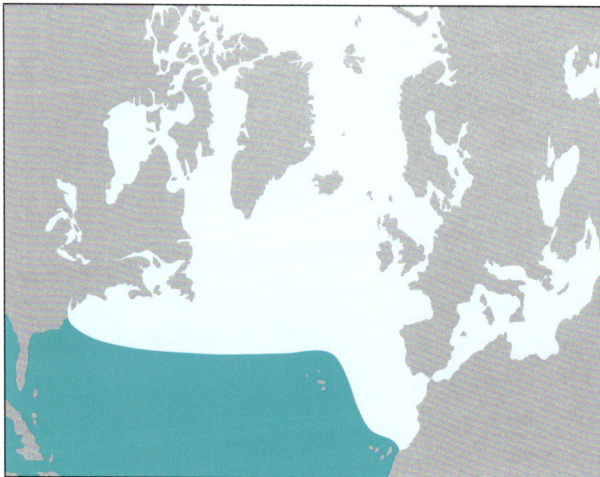

Distribution: Prefers the open ocean and the sea around oceanic islands. Comparatively common along the American east coast in the Gulf of Mexico. Occurrence in the eastern North Atlantic is very poorly documented, but the species can be expected in the southern part of the region covered by this book, perhaps at the Canary Islands. Seems not to occur so far north as the Atlantic Spotted Dolphin.

Behaviour: Pods vary in size from a few individuals to several thousand. Juveniles form small groups or join Spinner Dolphins. Groups may also consist of non-breeding mature females and immature males. Mother-calf pair is the most basic unit.

Movements apparently take place within a home range covering a linear distance of several hundred kilometres.

Swimming speed up to 40 km/hr. Dives last for up for 3–4 minutes; diving depth is not known but is probably a few hundred metres.

Occurs together with tunny (tuna fish).

Reproduction: There is no information from the northeast Atlantic, so the details reported must be treated with some reservation. Females reach sexual maturity at 9–11 years of age, males at 12–15 years. There are two mating periods: one in spring and one in autumn. Gestation lasts 11 months in this species. The precise duration of lactation is not known, but seems to vary between 1.4 and 2.1 years. Females normally give birth every third year.

Food: Small epipelagic fish.

Small pod of Pantropical Spotted Dolphins.

Natural enemies: Killer Whales and large shark species. Weakened individuals can be prone to attack by False Killer Whales.

Relationship with man: Accidental by-catches occur in bottom nets and drift nets.

Similar species: The Pantropical Spotted Dolphin is most often confused with the *Atlantic Spotted Dolphin*. The basic colour markings of the two species are, however, very different. The Pantropical Spotted Dolphin has a layered pattern, with dark cape, grey flanks and white belly, whereas the Atlantic Spotted Dolphin has a broken pattern in which the cape extends farther down and is absent behind the dorsal fin, this area being of the same colour as the flank.

Pantropical Spotted Dolphins which have few or no spots can be confused with the *Clymene Dolphin* or *Spinner Dolphin*, though they do not perform the spinning behaviour characteristic of those species. The Spinner Dolphin's very long beak also helps to rule out this species.

The two species of *common dolphin* are clearly differentiated by their colour markings.

Clymene Dolphin
Stenella clymene

ALTERNATIVE ENGLISH NAME Short-snouted Spinner Dolphin
FRENCH Dauphin de Clymene
GERMAN Clymene-Delphin
SPANISH Delfín clymene

Description: Comparatively stocky dolphin with medium-long beak and centrally positioned dorsal fin.
Size: Full-grown adults are 1.7–2 m long and weigh 50–90 kg; calves are c. 80 cm long and probably weigh c. 10 kg.
Teeth: There are 38–49 teeth in each half of both jaws.
Coloration: The Clymene Dolphin is tri-coloured, with black cape, grey flanks and white belly. Beak has black tip. Lips black. Seen from above, the nasal spot forms a 'helmet'. Has dark slender flippers.
Variation: Males are bigger than females. No differences between geographically distinct populations have been described.

Behaviour: Occurs in pods of 5–50 animals, but much larger groups of several hundred individuals are also known. The exact composition of groups is not known in detail.

Sometimes spins around its own axis. Bow-rides, but is rather shy in some areas. Is thought to forage at night.

Often seen with other dolphin species.

Reproduction: Almost nothing is known. It may be supposed that the details correspond fairly closely with those of other tropical and subtropical dolphins.

Food: Mainly fish and cephalopods.

Natural enemies: Presumably Killer Whales.

Pantropical Spotted Dolphin.

Clymene Dolphin.

Relationship with man: Accidental by-catches are reported. There is no commercial exploitation of this species.

Similar species: The Clymene Dolphin can be difficult to distinguish from the *Spinner Dolphin*, as the two have the same geographical range, behaviour (spinning leaps) and appearance. The Clymene Dolphin, however, has a much shorter snout which, when seen from above, reveals the unmistakable 'black helmet'. The Clymene Dolphin's dorsal fin is slightly more backward-curved than the Spinner Dolphin's.

At longer ranges, the Clymene Dolphin can also be confused with the other dolphin species in the region, but size, behaviour and beak length will quickly lead to a correct identification.

Distribution: Tropical and subtropical sea areas in the Atlantic Ocean. The distribution on the eastern side of the North Atlantic is very poorly known. The species may be expected at the Canary Islands. In the western North Atlantic, it is relatively common along the American east coast (Florida) and at the Bahamas and Bermuda, but there are reports all the way north to New Jersey.

Spinner Dolphin.

Spinner Dolphin
Stenella longirostris

FRENCH Dauphin longirostre
GERMAN Langschnauzen-Delphin
SPANISH Estenella giradora

Description: Slim, long-beaked dolphin with centrally positioned dorsal fin. Long, pointed flippers.
Size: Full-grown adults are 1.75–2.2 m long and weigh 40–80 kg; calves are c. 80 cm long and probably weigh c. 10 kg.

Teeth: There are 44–64 teeth in each half of both jaws.
Coloration: Tricoloured, with black cape, grey flanks and white belly.
Variation: Males are bigger than females. Any geographical variation in the North Atlantic has not yet been described.

Behaviour: Occurs in pods of between 5 and 200 animals. The groups are structured around family units with fluid social groups. Mother-calf pair is the basic unit. Groups consisting of a single sex or a single age class occur.

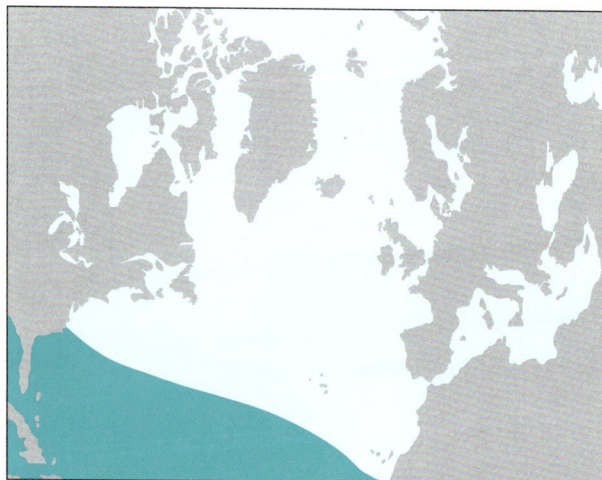

Distribution: Found in the North Atlantic south of 40°N, chiefly in oceanic seas and off oceanic islands. Along the USA's east coast it is found north to Cape Hatteras. On the eastern side of the North Atlantic there are reports from Cape Verde, and the Spinner Dolphin can be expected to appear occasionally at the Canary Islands.

The most characteristic type of behaviour is spinning, which is otherwise known only for the Clymene Dolphin. It can leap up to 3 m in the air, and it has been observed to spin 7 times in a single leap. It often rides the bow-waves of ships. Duration of dive is normally 1–2 minutes.

Forms schools with other dolphin species.

Reproduction: Age at sexual maturity is thought to be between 7 and 12 years for both sexes, but this has not been studied in any great detail in the North Atlantic. Animals can be mature from a length of 1.75 m. Mating is promiscuous and takes place from late spring to autumn.

Gestation lasts 10½ months, so that calving occurs in approximately the same period as mating. Lactation lasts 1–2 years. One calf is born every third year.

Food: Small fish and cephalopods.

Natural enemies: Probably Killer Whales and larger sharks. False Killer and Pygmy Killer Whales attack injured Spinner Dolphins which have escaped from fishing nets.

Relationship with man: Accidental by-catches in various types of fishing gear. No commercial exploitation. Atlantic Spinner Dolphins have not been kept in captivity.

Spinner Dolphins.

Similar species: The Spinner Dolphin is very difficult to distinguish from the *Clymene Dolphin*, with which it shares the same geographical range, behaviour (spinning leaps) and appearance. The Spinner Dolphin, however, has a much longer snout and its dorsal fin is slightly less backward-curved than the Clymene Dolphin's.

At longer ranges, the Spinner Dolphin could also be confused with the other dolphin species in the region, especially *Long-beaked Common Dolphin*, but colour markings and behaviour should quickly lead to identification.

Spinning leap performed by Spinner Dolphin.

Striped Dolphin.

Striped Dolphin
Stenella coeruleoalba

FRENCH Dauphin bleu et blanc
GERMAN Streifendelphin
SPANISH Estenella listada

Description: Slim dolphin with well-demarcated beak.
Size: Full-grown adults are 1.8–2.5 m long and weigh 110–170 kg; calves are c. 1 m long and weigh 20 kg.
Teeth: There are 39–55 teeth in each half of both jaws.
Coloration: Blue-grey back, grey flanks and white belly. At front of flanks a white or pale 'sideways V'. Distinct narrow black stripe from flipper to vent; at the flipper, this can be split into two or three stripes.
Variation: Males are slightly bigger than females. North Atlantic animals are bigger than Mediterranean ones. The stripe and the V can be weakly developed or lacking altogether on

Distribution: Compared with other *Stenella* species (spotted dolphins, Clymene Dolphin, Spinner Dolphin), lives farther north and in sea areas with greater temperature fluctuations. Is intermediate between Common Dolphin and the other *Stenella* species as regards its preference for oceanic habitats.

Normally does not occur north of 50°N, but in recent years it has been seen with increasing frequency along Scotland's north coasts, in the North Sea and in inner Danish waters. Recorded several times from the Norwegian coast north to 64°N, Iceland and the Faeroes. At Gibraltar it occurs in water depths of over 60 m, and in the eastern North Atlantic generally in relatively deep waters.

On the American east coast, the species is distributed north to Newfoundland. There is, moreover, a single record from west Greenland, though this may be regarded as a stray.

some individuals, making species identification difficult.

Behaviour: Pods are normally of fewer than 500 animals and around 100 on average. Groups of juveniles, groups of adults and mixed groups occur; the last two can be further divided into breeding groups and non-breeding groups. The young, after weaning, leave the adult group and join a juvenile group.

The behaviour at the surface is typical of dolphins, with leaps and bow-riding. The Striped Dolphin is thought to be capable of diving to several hundred metres and for up to 20 minutes, but detailed data are not available.

Mixed schools with other dolphin species are sometimes seen.

Reproduction: Both sexes become sexually mature at between 5 and 9 years of age. Mating and calving take place in spring, autumn and winter. Gestation lasts c. 12 months and lactation 18 months. One calf is born every third or fourth year.

Food: Cephalopods and fish.

Natural enemies: Killer Whales and large sharks.

Relationship with man: Accidental by-catches in various types of fishing gear occur. Commercial exploitation is not known in the region. Individuals have been kept in captivity on several occasions with a view to re-releasing them.

Similar species: Can be confused with most other dolphin species in the North Atlantic. Separated from the *Short-beaked Common Dolphin* by colour markings, from the *Long-beaked* also by beak length. Colour pattern is also a good distinguishing character from the *Spinner Dolphin*, *Clymene Dolphin* and the two *spotted dolphins*. *Fraser's Dolphin* has a stripe on the flank, but it is broader and differently positioned; in addition, Fraser's Dolphin has a very short snout.

Striped Dolphin leaping out of water.

Fraser's Dolphin.

Fraser's Dolphin
Lagenodelphis hosei

FRENCH Dauphin de Fraser
GERMAN Fraser Delphin
SPANISH Delfín de Fraser

Description: Comparatively stocky dolphin with very short snout and small centrally positioned dorsal fin, relatively long flippers.

Size: Full-grown adults are 2–2.7 m long and weigh 160–210 kg; calves are smaller than 1 m in length, with a weight of approximately 20 kg.

Teeth: There are 38–44 teeth in each half of both jaws.

Coloration: Black cape on back and grey flanks. White belly, with broad black stripe extending from base of beak to tail fluke; an offshoot of this runs from the corner of the mouth to the flipper. Flippers and tail flukes are similarly black.

Variation: Males bigger than females. No geographical differences yet described.

Distribution: Tropical and oceanic parts of the North Atlantic. Its occurrence seems sporadic, this due largely to insufficient information. The species is recorded from only two localities in the area covered by this book. It has been recorded around the Canary Islands, and it seems to occur slightly more frequently in the Gulf of Mexico.
In addition, a stranding on the Brittany coast has been recorded, but this was probably a stray pod.

Behaviour: Pods are large, with several hundred and sometimes several thousand animals. The exact composition of the groups is not known. Only limited behavioural information exists. Pods often mix with other species.

Reproduction: Only very incomplete information exists. Both sexes may be sexually mature at an age of 7–8 years and when 2.3 m long. The times of mating and calving are not known, nor is the period of gestation, though this is thought to be between 10 and 12 months.

Food: Fish and cephalopods.

Natural enemies: Presumably Killer Whales and large species of shark.

Relationship with man: Accidental by-catches in various types of fishing gear probably occur. No commercial exploitation of this species is known, nor has it been kept in captivity.

Similar species: The very short snout separates it from all other true dolphins in the region apart from the *White-beaked* and *White-sided Dolphins*, the normal distributions of which, however, do not overlap with that of Fraser's Dolphin. In addition, Fraser's Dolphin is smaller. It is the only dolphin which has a thick black stripe on the flank. On the *Striped Dolphin* the stripe is much narrower, and the *common dolphins* never show such a dark and broad stripe.

Fraser's Dolphin.

Rough-toothed Dolphin.

Rough-toothed dolphins

The rough-toothed dolphins represent a separate subfamily (Steninae) of the dolphins. They do not have a well-demarcated beak, and, as the name indicates, they have specially structured teeth. There are four species. The Tucuxi Dolphin (*Sotalia fluviatilis*) occurs along the northeast coast of South America and in the Amazon and its larger tributaries, thus outside the area covered by this book. The Indopacific Humpback Dolphin (*Sousa chinensis*) lives in the Indian Ocean and the western Pacific Ocean, but occasional individuals have strayed through the Suez Canal to the Mediterranean Sea. The other two species in the subfamily occur in the North Atlantic: the Rough-toothed Dolphin (*Steno bredanensis*), which is a widely distributed, tropical–subtropical oceanic species, and the Atlantic Humpback Dolphin (*Sousa teuszi*), which is a regular coastal species and is found off West Africa.

Rough-toothed Dolphin
Steno bredanensis

FRENCH Sténo
GERMAN Rauzahn-Delphin
SPANISH Esteno

Description: Slim dolphin with a head that has been called 'reptilian'. The beak is not well defined, but tapers gradually outwards from the melon. The dorsal fin is centrally positioned.
Size: Full-grown adults are 2.1–2.8 m long and weigh 100–150 kg; length at birth probably c. 1 m, weight at birth not known.
Teeth: 20–27 teeth in each half of both jaws, teeth with a characteristic rough upper surface.
Coloration: Narrow, dark grey cape, paler grey on the flanks and white on the belly. Many animals have white scars on the body caused by squid and sharks.
Variation: Males and females are alike in size. No geographical differences have been described.

Behaviour: Groups of between 10 and 20 are most common, but larger groups of up to 100 occur. Their exact composition is not known.

Often makes shallow leaps. Sometimes rides the bow-waves of ships, but this

Distribution: Open tropical seas. Normally not farther north than 40°N. Does, however, occur in Bay of Biscay, around Madeira and in the Mediterranean Sea.

On the American east coast it is known from Florida in the south to Cape Hatteras in the north, but records close to the coast involve stray individuals. Main area of occurrence is in deeper parts of the North Atlantic.

behaviour is seen more rarely in this species than in the other tropical dolphins. The Rough-toothed Dolphin often travels at high speed with its head out of the water, but it can also spend a long time, c. 15 minutes, underwater.

Sometimes seen with Bottlenosed Dolphins and other dolphins, and also with pilot whales.

Reproduction: Females reach sexual maturity at a length of 2.1 m, males at around 2.25 m. Their respective ages at maturity are 10 and 14 years, assuming that the North Atlantic populations are similar to the Japanese ones. It is not known when mating and calving take place, nor how long the lactation period is. Gestation is between 10 and 12 months in duration.

Rough-toothed Dolphins.

Food: Fish and cephalopods.

Natural enemies: Killer Whales and larger sharks.

Relationship with man: There is no commercial hunting of this species, but accidental by-catches in certain types of fishing gear occur. Animals have occasionally been kept in captivity, and in one instance a hybrid young was produced from a crossing between a Bottlenosed Dolphin and a Rough-toothed Dolphin.

Similar species: The head shape and the distinct white scars are normally distinctive of the Rough-toothed Dolphin. At longer ranges, however, it can resemble the *Bottlenosed Dolphin* and other tropical dolphins. *See also* the comparative plate on page 42.

Atlantic Humpback Dolphin.

Atlantic Humpback Dolphin
Sousa teuszi

FRENCH Dauphin à bosse de l'Atlantique
GERMAN Atlantischer-Buckeldelphin
SPANISH Delfín jorobado

Description: Robust dolphin, with dorsal fin characteristically set on top of a 'hump'. The beak is not well demarcated, but tapers gradually from the head.
Size: Full-grown adults are 2–2.8 m long and weigh 100–284 kg; young are 1 m long at birth, weight at birth not known but estimated at 10–12 kg.

Teeth: There are 26–31 teeth in each half of both jaws.
Coloration: The back is typically ash-grey to light grey and the belly white.
Variation: Males are bigger than females. Coloration varies somewhat, both very dark and very pale individuals occurring. The animals generally become darker with age; the dorsal fin, however, becomes paler. The shape of the dorsal fin varies as well. The hump can be absent on young animals, and develops with age.

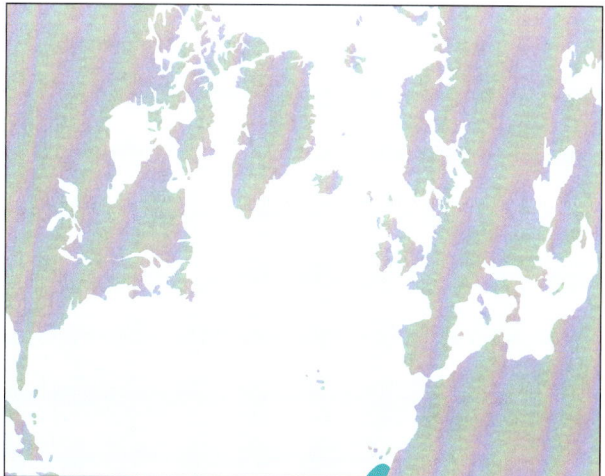

Distribution: Coastal waters and river mouths along the African west coast and north to the Canary Islands.

Atlantic Humpback Dolphin reveals why it is so named.

Behaviour: Groups of 5–7 animals, sometimes up to 25, are seen. The group structure is not known.

Does not ride the bow-waves of ships and avoids vessels. Spyhopping has been recorded. Sometimes swims on its side and waves its flippers. Comes to the surface every 40–60 seconds, but can also remain submerged for several minutes.

Sometimes seen with Bottlenosed Dolphins.

Reproduction: Very poorly known. Neither length nor age at maturity is known. Limited observations suggest a calving period in the early months of the year. Gestation period is probably between 10 and 12 months.

Food: Fish and cephalopods.

Natural enemies: Possibly Killer Whales.

Relationship with man: There is no commercial hunting of this species, but accidental by-catches may occur. It has not been kept in captivity.

Similar species: Because of the characteristic hump, this species is not normally confusable with other dolphins. Individuals with a poorly developed hump can, however, resemble the *Bottlenosed Dolphin*.

Note: The Indopacific Humpback Dolphin (*Sousa chinensis*), which occurs in the Red Sea, has been known to swim up the Suez Canal to Port Said and the Mediterranean coast of Israel. The possibility of this species appearing in the eastern Mediterranean cannot, therefore, be ruled out. Such cases involve human-influenced, not natural, immigration.

Blackfish
and killer whales

This subfamily (Orcinae) contains small to medium-sized species having a rounded head without a well-defined beak. They can be divided into three groups: 'blackfish', killer whales and Risso's dolphin. The last-named is possibly more closely related to the true dolphins, but is nevertheless, for practical reasons, described here.

The genera *Globicephala*, *Pseudorca*, *Peponocephala* and *Feresa* are often referred to collectively as 'blackfish', on account of their predominantly black colour.

Comparative plate of blackfish.

Melon-headed Whale, page 79

Long-finned Pilot Whale, page 74

Pygmy Killer Whale, page 80

Short-finned Pilot Whale, page 77

False Killer Whale, page 82

Long-finned Pilot Whale.

Long-finned Pilot Whale
Globicephala melas

FRENCH Globicéphale noir
GERMAN Langflossen Grindwal
SPANISH Calderon común

Description: Medium-sized cetacean with rounded head and very long, pointed flippers, the length of which corresponds to 18–27% of the body length. The mouth is slightly upturned. The dorsal fin is set typically on the front third of the back and is broad-based.

Size: Full-grown adults are 3.8–6 m long and weigh 1.8–3.5 tonnes; calves are 1.8–2 m and weigh c. 75 kg.
Teeth: There are 8–13 teeth in each half of both jaws; they are set in the front third of the lower jaw and fill up only the first half of the dental strip.
Coloration: Black on back and flanks, sometimes with a greyish area behind the dorsal fin. The cape is sometimes more strongly demarcated, and the back then looks darker than the flanks. An anchor-shaped patch on the belly.

Distribution: Oceanic parts of the North Atlantic, from Greenland and Norwegian coasts in the north, south to the Canary Islands on the eastern side and the Virginian coast on the western side. Common in the western Mediterranean Sea. Known especially from the seas around the Faeroes. Occasional in the North Sea.

Variation: Males bigger than females; on older males, the size of the head is particularly distinctive. On younger animals, the dorsal fin is set closer to the middle of the back.

Behaviour: Groups of thousands of animals are known to occur. These are composed of several generations of females along with young of both sexes. Adult males in the group do not breed within it, but apparently mate with females in other groups.

The blow is sometimes visible and up to 1 m tall. When foraging, dives for up to 10 minutes. Can dive to 600 m, but mostly it descends to only 30–60 m.

Long-finned Pilot Whales.

Reproduction: Males reach sexual maturity at a length of 4.9 m, females between 3.66 m and 4.27 m. The age at the onset of maturity is on average 12–16 years for males and 6–9 years for females. Gestation lasts probably 12–16 months. The lactation period can last for 3–4 years, and is sometimes further protracted in older females which are no longer reproductively active. The female produces on average one calf every fifth year.

Food: Cephalopods, crustaceans and shoal fish.

Natural enemies: There are reports of attacks by Killer Whales on injured pilot whales in Bay of Biscay. In other localities, however, Killer Whales and pilot whales seem to avoid each other.

Relationship with man: Hunting of Long-finned Pilot Whales is or has been carried out in many places in the North Atlantic. Whaling stations in Newfoundland, Iceland and the Shetland Isles have closed down, and the only one remaining is in the Faeroes. During the local Faeroese hunt, between 1,000 and 2,000 pilot whales are killed annually. The pilot whales are utilised on an exclusively local basis. A comprehensive study of their population levels has confirmed that the population is not threatened by this exploitation.

To some extent, accidental by-catches in various types of fishing gear also occur.

Similar species: The only other species with a similar rounded head is the *Short-finned Pilot Whale*. Where the two species overlap in range, it is not possible to distinguish between them with certainty.

With stranded animals, the Short-finned Pilot Whale's smaller number of teeth and shorter flippers may determine which species is involved, but there is overlap between the two in both characters. *See also* the comparative plate on page 73.

Short-finned Pilot Whale.

Short-finned Pilot Whale
Globicephala macrorhynchus

FRENCH Globicéphale tropical
GERMAN Kurzflossen Grindwal
SPANISH Calderon de aletas cortas

Description: Medium-sized cetacean with rounded head and longish flippers, though the latter, in contrast to those of its close relative, have a length corresponding to only 14–19% of the animal's total length. The dorsal fin is set typically on the front third of the back.
Size: Full-grown adults are 3.6–6.5 m long and weigh 1–4 tonnes; calves are 1.4–1.9 m long and weigh c. 60 kg.

Teeth: There are 7–9 teeth in each half of both jaws.
Coloration: Looks uniformly black at a distance, but at close range different shades can be seen. There is a darker cape and paler flanks, while an anchor-shaped patch is present on the belly. Greyish or much paler areas are often present behind the dorsal fin.
Variation: Males grow larger than females, and with age they develop a bigger and more swollen head which protrudes beyond the poorly demarcated beak. On young animals, the dorsal fin is more centrally placed. Young animals, probably because of their smaller size, leap right out of the water more often than do adults.

Distribution: There are occasional reports from Bay of Biscay and the French coast. A resident population is present around the Canary Islands. On the western side of the North Atlantic, the species occurs north to the Virginian coast. The northern limit of the Short-finned Pilot Whale and the southern limit of the Long-finned Pilot Whale overlap each other.

Behaviour: Occurs in groups varying greatly in size, from 10–30 to several thousand animals. The group structure is probably similar to that of the Long-finned Pilot Whale, consisting of female groups with young of both sexes. Adult males are sometimes included in groups, but DNA studies have shown that they do not breed within their own group.

Undertakes migrations, the extent of which is not known, but which probably involve diffuse, nomadic movements. Some populations, such as that around the Canaries, remain in the same area all year and are therefore non-migratory.

From time to time an entire group can be seen logging at the surface, allowing vessels to approach very closely. Tail-slapping and spyhopping are observed. Presumably forages mostly at night. Mass strandings of this species are known, but the causes of these are as yet poorly understood.

Can probably dive to several hundred metres for cephalopods. Dives typically last for 10 minutes, but longer dive times are known.

Reproduction: Data from the North Atlantic are scant, but the details are probably similar to those prevailing in other parts of the range, where males reach sexual maturity at an age of 12 years and a length of 5 m. Females can be sexually mature when they are 3.5 m long. Females cease producing young at an age of 35 years, but they continue to suckle calves in the group (probably closely related animals).

Food: Cephalopods, at times also shoal fish.

Natural enemies: Probably Killer Whales.

Relationship with man: Commercial hunting takes place in the Caribbean. It is probably subject to accidental by-catches in fishing gear. The species is the object of extensive whale tourism around the Canary Islands, which are the world's third biggest whale-watching locality.

Similar species: At sea, the two species of pilot whale cannot be distinguished from each other with certainty. Normally, geographical exclusion will determine which species is involved. In the overlap zone, however, it is difficult, but the Short-finned species normally has a paler back than the Long-finned.

Dead animals can be identified to species by the number of teeth and the length of the flippers, but even here there is overlap between the two species.

See also the comparative plate, page 73.

Short-finned Pilot Whales.

Melon-headed Whale.

Melon-headed Whale
Peponocephala electra

FRENCH Péponocéphale
GERMAN Melonenkopf
SPANISH Calderon pequeño

Description: Small black whale with oval head. The flippers are comparatively long (one-fifth of body length) and very pointed.
Size: Full-grown adults are 2.1–2.8 m long and weigh 160–210 kg; calves are c. 1 m long, with weight at birth unknown but presumably c. 10 kg.
Teeth: There are 20–26 teeth in each half of both jaws.
Coloration: Almost black over the whole body, though the lips are sometimes white. Belly palest, and on the back it has a cape, which is broadest below the dorsal fin.
Variation: No difference in size between the sexes.

Distribution: The Melon-headed Whale occurs in the world's tropical, subtropical and warm-temperate seas. In the North Atlantic it is recorded relatively frequently on the western side, namely in the Gulf of Mexico and the Caribbean Sea. On the eastern side, there are just a few reports from the Senegal coast and off Sierra Leone and the Cape Verdes. In Europe, it has stranded only once (a single individual), on the Cornish coast of England. Apparently, it occasionally migrates farther north with the Gulf Stream, which would explain this stranding. This species should be looked for around the Canary Islands and along the northwest African coast.

Behaviour: Lives in groups of between 150 and 1,500 animals. The groups are often very tightly packed, and travel rapidly with frequent changes of direction. The sex ratio is known from a group which stranded, in which there was a preponderance of females. If this division reflects the usual situation, there may be a degree of separation between the sexes.

Rides the bow-waves of ships and makes low, shallow leaps out of the water when swimming rapidly. Spyhopping is also known. When travelling more slowly, the head is stuck out of the water. Is normally shy of stationary boats or slow-moving vessels.

Reproduction: Males become sexually mature at an age of between 3 and 7 years, whereas females are 4–12 years old at maturity. The smallest sexually mature male recorded was 2.48 m long, the smallest mature female 2.3 m. The gestation period is presumably 12 months. Length of lactation period is not known.

Food: The diet consists of shoal fish and larger species of cephalopod.

Natural enemies: Probably Killer Whales.

Relationship with man: Accidental by-catches in fishing gear evidently occur.

Similar species: The Melon-headed Whale can easily be confused with the *Pygmy Killer Whale*, which is of the same size and broadly the same coloration. The Melon-headed Whale, however, most often occurs in larger groups and it mixes with other species. The shape of the head seen from above is more triangular, and the dorsal cape is broader below the fin and extends farther down. The tip of the flippers is pointed, but this characteristic is difficult to see in the field. When stranded, it is identified by the large number of teeth.

Pygmy Killer Whale
Feresa attenuata

FRENCH Orque pygmée
GERMAN Zwergschwertwal
SPANISH Orca pigmea

Description: Small, rather slim whale with a centrally placed dorsal fin. The flippers are comparatively short and broad, and rounded at the tip.
Size: Full-grown adults are 2–2.7 m long and weigh 110–170 kg; calves measure around 80 cm in length and probably weigh c. 10 kg.
Teeth: There are 8–13 powerful teeth in each half of both jaws.
Coloration: Overall, the entire animal appears to be uniformly dark grey, but the cape can be distinguished; in side-on view, this is narrow at the front and broad below the dorsal fin. The lips are often white, and a white patch of variable extent is visible roughly in the middle of the belly.
Variation: Relatively little sexual dimorphism: males grow to 2.7 m long, females to marginally less.

Behaviour: Group size is between 15 and 50 animals.

Intermittently sticks its head right out of the water. Also makes dolphin leaps. Mass strandings occur. Is regarded as rather shy of ships and is therefore difficult to approach closely. Logging at the surface, spyhopping and various tail-slapping are seen. Is not very active.

Mixed groups with False Killer Whales and Melon-headed Whales are reported.

Reproduction: Very poorly known. Females become sexually mature when c. 2 m long, males presumably at 10 cm longer. Gestation and lactation periods are not known, but are assumed to be similar in duration to those of other small cetaceans.

Food: Fish, but there are unconfirmed

Pygmy Killer Whale.

reports that the Pygmy Killer Whale is also capable of taking much bigger prey, including dolphins.

Natural enemies: Possibly Killer Whales.

Relationship with man: The species is not hunted. Accidental by-catches in fishing gear probably occur locally. Elsewhere in the world, has been kept in captivity for brief periods.

Similar species: Other blackfish can be confused with the Pygmy Killer Whale. Species identification is not made any easier by the fact that they appear in mixed groups. With dead animals, size, the shape of the flipper ends, the extent of the cape and, of course, the number of teeth can facilitate identification. *See also* the comparative plate on page 73.

Distribution: An oceanic warm-water species. Recorded in Bay of Biscay, but could also be expected to occur along the northern offshoot of the Gulf Stream and perhaps in the western Mediterranean, as well as on the West African coast. There are few observations off the American east coast. The available knowledge on this species from the North Atlantic stems almost exclusively from the Caribbean Sea.

False Killer Whale.

False Killer Whale
Pseudorca crassidens

FRENCH Faux orque
GERMAN Kleiner Schwertwal
SPANISH Falsa orca

Description: Medium-sized whale with dorsal fin on centre of back. The flippers are one-tenth the length of the body and are very characteristic, having a distinct kink or elbow on the front edge.

Size: Full-grown adults are 4–6 m long and weigh 1,100–2,200 kg; calves are 1.6–1.9 m long, with a weight of c. 80 kg.
Teeth: There are 8–11 powerful teeth in each half of both jaws.
Coloration: Like the other blackfish, the False Killer Whale is at first glance almost entirely black. On closer inspection, however, it is possible to make out a black cape and a dark grey flank. Some individuals have white areas on the underside.

Distribution: The False Killer Whale is an oceanic warm-water species the natural range of which, in the North Atlantic, lies south of 40°N. Groups can, however, travel farther north with the Gulf Stream. In some years, it is encountered in the North Sea and even as far north as the Faeroes. The species was believed extinct until 1861, when a large group turned up almost 'out of the blue' in the western Baltic and the Danish straits.

Occurs commonly in the western Mediterranean, but has not been confirmed in the eastern Mediterranean and the Black Sea.

Along the American east coast, it has been recorded from Florida in the south to Cape Cod in the north.

Variation: Males are generally bigger than females of the same age. Young animals can be paler than adults. The tip of the dorsal fin can vary somewhat among individuals: some animals have a rounded tip instead of the more common sail-shaped dorsal fin.

Behaviour: Groups are typically of between 30 and 50 animals, but much bigger gatherings also occur. It is not known whether the group structure is more like that of the Killer Whale or like that of the pilot whales.

Fast and active swimmer. Often raises the head and forebody out of the water when submerging. Sometimes only the characteristic flippers are visible. The mouth is frequently open, so that the teeth can be seen. Spyhopping and bow-riding are known. The species is not shy and quite often comes up to boats to inspect them. Often leaps clear of the water.

A number of mass strandings have occurred. Diving behaviour is not known in detail. It has been estimated that this species can dive to around 500 m.

False Killer Whales are sometimes seen swimming with other dolphin species, e.g. the Short-finned Pilot Whale.

Reproduction: Males become sexually mature when c. 4 m long and at an age of 8 years, females at c. 3.5 m and likewise at

False Killer Whales.

8 years. Timings of mating and calving are not known, but these perhaps take place all year around. Gestation lasts 16 months, and normally just one calf is born. The lactation period is estimated at 18 months. The number of young produced per female is not known.

Food: Eats fish and cephalopods, but is also known to attack wounded dolphins which have escaped from tuna nets.

Natural enemies: Possibly Killer Whales and, in the case of young, perhaps large sharks. There are, however, no reports of such attacks.

Relationship with man: Accidental by-catches represent the greatest human-induced cause of death. Both in Europe and in the USA, False Killer Whales are, or have been, kept in captivity. Hybridisation between captive False Killer Whales and Bottlenosed Dolphins has been reported.

Similar species: The False Killer Whale can in principle be confused with all other blackfish. From *pilot whales* it is distinguished by the head and flipper shapes, if these are visible, as well as by the general behaviour, as it is far more active than the more 'lethargic' pilot whales.

From the *Pygmy Killer Whale* and *Melon-headed Whale* it is distinguished primarily by size and coloration, but also by the appearance of the flippers. In the case of stranded animals, it is easy to determine which species is involved by counting the number of teeth.

See also the comparative plate on page 73.

Killer Whale
Orcinus orca

ALTERNATIVE ENGLISH NAME Orca
FRENCH Orque
GERMAN Schwertwal
SPANISH Orca

Description: Medium-sized whale without well-demarcated beak, but with centrally positioned dorsal fin.
Size: Full-grown adults are 4.6–9.8 m long and weigh 2.6–9 tonnes; calves are c. 2.3 m long and weigh c. 180 kg.
Teeth: There are 10–14 powerful teeth in each half of both jaws; the teeth are oval in cross-section.
Coloration: The back, most of the flank and the flippers are black. A white patch behind the upper level of the eye, and a greyish 'comma patch' behind the dorsal fin. A 'white finger' pointing backwards on the rearmost third of the flank. Most of the belly, as well as the underside of the tail flukes, is white. There is a black patch around the genital openings.
Variation: Males grow considerably bigger than females (almost 10 m long and 9 tonnes in weight, whereas females reach a length of only 7 m and a weight of 3 tonnes). The adult male's dorsal fin is sword-shaped and up to 2 m tall; it is strikingly visible.

Behaviour: Killer Whales live in groups of 10–50 animals, all of which are related to one another. Each group has its own dialect of 'Killer Whale language'.

These whales hunt in groups and have developed a large number of hunting techniques. They may strand themselves in order to snatch a resting seal, and then wriggle their way back out to sea again. They can also use sounds to concentrate a fish shoal into a solid column, which is easy to attack. The Killer Whale can leap right out of the water. Spyhopping occurs frequently. Short dives last 10–35 seconds, longer ones up to 4 minutes, with 17 minutes the longest documented dive time. The Killer Whale can dive to at least 260 m.

Killer Whale, male.

Killer Whale, female.

Distribution: The Killer Whale is both an oceanic and a coastal species. In the North Atlantic, it can in principle, therefore, be seen anywhere. Even though it should be described as cosmopolitan, it is associated mostly with the colder sea areas.

In the eastern part of the North Atlantic, it is quite common off Iceland and at several places along Norway's west coast. It appears frequently in the Skagerrak and is often seen on Scottish coasts. Killer Whales are also seen frequently in the Irish Sea and Bay of Biscay. In the Mediterranean, the species is most frequent in the western part and is relatively rare in the eastern part. Killer Whales have not been documented with certainty in the Black Sea.

In the western North Atlantic, the Killer Whale is known from west Greenland and Canadian coasts. Farther south, along the east USA coast, the species is seen especially in the Gulf of Maine, but its distribution presumably extends right down to Cape Hatteras. South of Cape Hatteras, it is rarely seen. Records at the Florida coast may originate from populations found in the Gulf of Mexico.

This is one of the fastest of the cetaceans, with a top speed of over 60 km/hr.

Reproduction: Males become sexually mature when 5.2 m long and at an age of 16 years, while females can be sexually mature at 4.6 m and 8 years of age. Gestation lasts 19 months, and lactation can extend over more than 1 year. Reproduction takes place extremely slowly in Killer Whales, as only mature females of high rank give birth to young. There is therefore, on average, only one young born every fifth year.

Food: The Killer Whale's diet includes a large number of fish and cephalopods, but besides these it is known to eat seals, porpoises, dolphins and seabirds. Even large cetaceans such as the Blue and Sperm Whales are on its menu.

Killer Whale pod.

Killer Whale surfacing.

Natural enemies: None.

Relationship with man: Until the 1970s, commercial hunting of Killer Whales was carried out on a small scale by Norway and Iceland. In addition, animals were captured and taken into captivity from Icelandic and other waters right up to the 1980s.

Similar species: Because of its size and coloration, as well as its spectacular behaviour, the Killer Whale cannot normally be confused with other species. The males' dorsal fin, the white patch behind the eye and the grey comma-mark behind the dorsal fin are unmistakable.

The *White-beaked Dolphin* is the only other cetacean with sharply defined black and white markings, but it is much smaller and is characterised by its much more acrobatic behaviour.

Killer Whale spyhopping.

Risso's Dolphin.

Risso's Dolphin
Grampus griseus

FRENCH Dauphin de Risso
GERMAN Rundkopf-Delphin
SPANISH Delfín de Risso

Description: Medium-sized dolphin with rounded head and characteristic forehead crease (*see* figure below). Tall, centrally placed dorsal fin.
Size: Full-grown adults are 2.6–4 m long and weigh 275–600 kg; calves are c. 1.2 m long and presumably weigh c. 20 kg.
Teeth: Teeth are present only in the lower jaw, where there are 4–7 on each side.
Coloration: Typically, medium to dark grey on the back and paler on the flanks, with a well-marked border with the white belly. This colour pattern is often concealed behind scars made by squid, and at times the underlying coloration can be more or less obliterated by these.
Variation: Males are somewhat bigger than females. Old animals, especially old males, can be almost entirely white.

Forehead crease on Risso's Dolphin.

Behaviour: The pods consist typically of between 3 and 50 animals, but larger groups are observed. Group structure is not known.

Leaps up from water like other dolphins. Spyhops by pushing its forebody well out of the water. Seldom bow-rides. Typically dives for 1–2 minutes, but can remain submerged for as long as 30 minutes.

Reproduction: Not particularly well known. The animals become sexually mature when between 2.4 and 3 m long. Age at onset of maturity is not known. Gestation presumed to last 13–14 months. Duration of lactation is not known.

Risso's Dolphin spyhopping.

Pod of Risso's Dolphins.

Food: Cephalopods.

Natural enemies: Probably Killer Whales and large shark species.

Relationship with man: There is no commercial hunting of the species, but, as most other small cetaceans, Risso's Dolphin is subject to accidental by-catches in fishing gear. Individuals have occasionally been kept in captivity.

Similar species: The scarred appearance with bright white areas makes Risso's Dolphin normally impossible to confuse with other species. Old individuals, however, can be almost completely white and are therefore confusable with the *Beluga*, although the distributions of these two species only occasionally overlap; in any case, the tall dorsal fin at once distinguishes it from the Beluga. In head-on view at close range, the forehead crease is unmistakable. Younger animals can resemble the *Bottlenosed Dolphin* in colour, but the head shape distinguishes them.

Distribution: Preferred habitat is the subtropical and warm-temperate oceanic seas, but in some places it occurs close inshore.

In the North Atlantic, it is comparatively frequent in the Irish Sea and Bay of Biscay. The northernmost record is from Trondheim Fjord. Has been found a few times in the North Sea and the Baltic Sea. Quite frequent in the Mediterranean.

Along the American east coast, the species is known from as far north as Newfoundland. Reports from Greenland are, however, erroneous.

Porpoises

The 6 species in the porpoise family (Pho-coenidae) are closely related to the members of the dolphin family. They were earlier included as a subfamily of the latter. The porpoises differ, however, from the dolphins in having spatulate, not conical, teeth and in a number of other anatomical and structural characteristics. Porpoises are chiefly coastal and prefer temperate seas in the Northern and Southern Hemispheres.

Only a single species, the Harbour Porpoise (*Phocoena phocoena*), lives in the North Atlantic.

Harbour Porpoise
Phocoena phocoena

FRENCH Marsouin commun
GERMAN Schweinswal, Kleintümmler
SPANISH Marsopa común

Description: Small, compact cetacean without well-defined beak. On the centre of the back there is a small, triangular dorsal fin up to 20 cm tall, the front edge of which has skin nodules, or tubercles.
Size: Full-grown adults are typically 1.3–1.8 m long and weigh 50–90 kg, though animals of over 2 m have been recorded several times; calves are c. 75 cm long and weigh 8–10 kg.

Spatulate teeth of Harbour Porpoise.

Teeth: 23–28 spatulate teeth in each half of both jaws.
Coloration: The back is bluish-black and the belly white. On the foreflanks there are grey areas of variable size and extent. A dark stripe runs from the corner of the mouth to the leading edge of the flipper.
Variation: Females are a little bigger than males. In some regions young animals have a darker coloration than adults, with a diffuse dark stripe along the centre of the belly. There are geographical differences in the frequency of tubercles; e.g. the Black Sea population has far fewer tubercles than the Atlantic population. The grey panel on the forebody can encompass the upper lips and parts of the head, so that the entire forebody looks grey.

Behaviour: Groups of up to 10 animals occur, but these involve loose groupings. The basic element is mother-calf pair, which sometimes gather in larger units.

There are differences in the extent of the sea areas in which the different age groups and sexes live. Young, juvenile animals roam over greater distances than do adults. Pregnant females return each year to the same 'small' sea area, where they remain throughout the summer, and where births

Harbour Porpoise.

Two Harbour Porpoises.

and subsequent matings take place. Sexually active adult males travel between these breeding areas. Seasonal movements have been described from the Baltic and the Black Seas, with immigration in spring, and emigration in late autumn and winter as the ice starts to form. In other places, movements to and from the coast take place.

In some places the Harbour Porpoise is described as a shy animal, while in others it is frequently attracted to yachts and slow-moving motorboats. It can leap right out of the water and will bow-ride, but is best known for its rolling actions. Dive times can be up to 6–8 minutes, but are normally only c. 3 minutes. The maximum depth of dive is probably 200 m, but the animals do not normally dive this deep.

Reproduction: Males become sexually mature at a length of 1.35 m and an

Distribution: The Harbour Porpoise's preferred habitat is coastal and shallow offshore waters. In the northeast Atlantic, it occurs along coasts from the White Sea in the north to West Africa in the south. There are large populations in the North Sea and Danish waters. In the inner Baltic and the English Channel the porpoise is now rather rare, but the Irish and Celtic Seas hold large populations. A population was formerly found in the west Mediterranean Sea, and observations on the Tunisian coast may indicate the presence of a small relict population there. An isolated population is found in the Black Sea, from where there are occasional movements into the Aegean Sea via the Bosporus, the Sea of Marmara and the Dardanelles. The Harbour Porpoise is common along the Norwegian coast and also off Scotland, the Faeroes and Iceland.

The Harbour Porpoise is also documented from various European rivers such as the Elbe, the Rhine, the Thames, and the Neva River between Russia and Estonia.

In the western North Atlantic, the species occurs along the American east coast south to North Carolina (Cape Hatteras). The main distribution, however, is thought to be in the Gulf of Maine, Bay of Fundy and Gulf of St Lawrence. The Labrador and Newfoundland coasts are other important regions for the species in the western North Atlantic. On east Greenland coasts it is only a rare visitor, whereas it has a significant presence off west Greenland.

age of 2–3 years, females at 1.45 m and at 3–4 years. Matings take place in late summer. Gestation lasts c. 11 months, and births therefore take place in the following summer. The lactation period is 5–8 months. An adult female can therefore produce one young each year. In reality, the frequency of calving is somewhere between every year and every other year.

Food: The Harbour Porpoise's diet contains a wide variety of fish and cephalopods. Detailed studies undertaken in several places in the North Atlantic indicate that this species is rather flexible in its choice of food. In the northernmost part of the range it feeds predominantly on capelin, while farther south this is supplemented with herring and codfish. In the Black Sea, gobies are an important element in the diet.

Natural enemies: Killer Whales and large sharks are documented as natural enemies. The stomach of a single Killer Whale stranded at Grenå, Denmark, in 1861 contained, besides a number of Common Seals, the remains of no fewer than 13 Harbour Porpoises.

Relationship with man: The Harbour Porpoise has been commercially exploited in several parts of Europe. Its annual migrations into and out of the Baltic were regularly exploited; in the 1800s, at least 140,000 porpoises were killed in Danish waters. Large-scale hunting of the species is still carried out in the Black Sea, and this has led to a serious reduction in the population. In Greenland, between 500 and 1,000 porpoises are killed annually.

The greatest mortality, however, occurs in the form of by-catches in subsurface and surface gill nets throughout the range. North Sea fishermen alone capture between 5,000 and 7,000 animals annually in their nets, but there is also a large by-catch off the North American east coast.

The Harbour Porpoise is kept in captivity in a few places. This includes individuals kept for scientific research aimed at solving the problems of by-catches.

Similar species: The Harbour Porpoise's tiny size and small triangular dorsal fin make it relatively easy to identify. The characteristic roll is also a good identification aid.

Harbour Porpoises' prelude to mating.

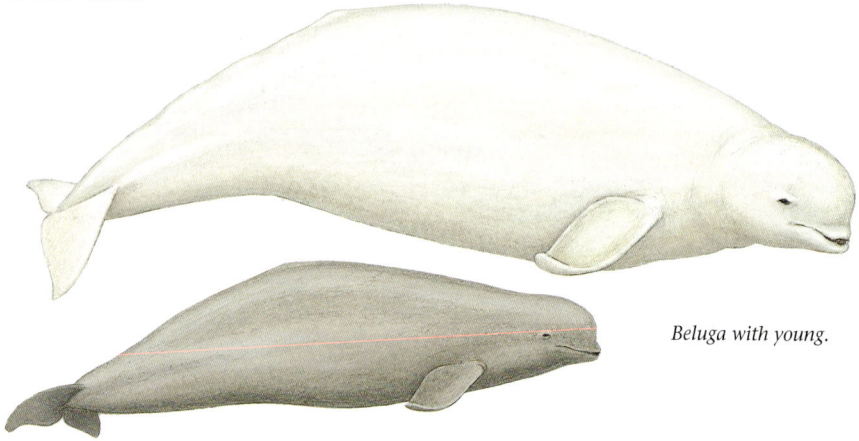

Beluga with young.

Beluga and Narwhal

This small family (Monodontidae) consists of just two species, which are closely
related to both the dolphins and the porpoises. The two are true arctic whales: the
Beluga (*Delphinapterus leucas*) and the Narwhal (*Monodon monoceros*). Both species
exhibit strong specialisation in dentition,
have evolved mobile flippers, and lack a
dorsal fin. Hybridisation between the two
has been described from Greenland.

Beluga
Delphinapterus leucas

ALTERNATIVE ENGLISH NAME White Whale
FRENCH Porpoise blanc, Belouga
GERMAN Weißwal, Beluga
SPANISH Belugua

Description: Compact, medium-sized
whale with highly mobile head. The lips
can be formed into various shapes and
produce a rich repertoire of sounds. There
is no dorsal fin, but instead a swelling in
the centre of the back. The flippers are
mobile, and are often curved upwards
along the leading edge.
Size: Full-grown adults are 3–5 m long and
weigh 400–1,500 kg; calves are 1.5–1.6 m
long, with a weight of c. 80 kg.

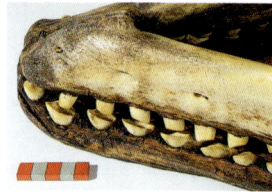

Beluga's characteristic worn teeth. The jaws are shown from below.

Teeth: There are 8–11 teeth in each half of
both jaws; typical of the Beluga, these often reveal signs of wear (*see* photograph),
as a result of which the teeth of the upper
jaw may become hollowed out.
Coloration: The entire body of adults is
brilliant white.
Variation: Males are generally bigger than
females. Coloration changes with age.
Young animals are ash-grey, but a gradual
and patchy transition from ash-grey to
bright white takes place with age. Juveniles
can thus have large areas of both grey and
white on the body. Only adults are all
white.

Behaviour: Groups of between 5 and 10
animals, but gatherings of up to several
thousands are known, especially in river
mouths. Group structure is not known,
but is probably not so complex as that of
some dolphins. The basic element seems
to be mother-calf pair.
  Spyhopping and tail-slapping are common. On the whole, does not leap out of

the water but raises the head above the surface while swimming. A slow-swimming species. Spends a significant part of the day at the surface. Diving sequence consists of 5–6 shallow dives in the course of a minute, followed by deeper and longer dives. In deeper water, dives to 600 m and of up to 15 minutes' duration have been recorded.

Since their sonar is adapted to coastal conditions, Belugas rarely strand themselves, unless accidents or by-catches are involved. The Beluga is sometimes caught out by the tide and accidentally ends up on a sandbank. This does not seem to cause the animals stress, as they calmly swim away at the next high water.

Distribution: Arctic, inshore sea areas, but is also able to survive in coastal waters farther south.

Not uncommon on Iceland's north coast. Stray individuals sometimes swim down the Norwegian coast, and occasionally reach the North Sea and the Baltic Sea (several times during the 1980s and 1990s).

Migrates up rivers. A Beluga was seen in the Rhine in 1967, and another in the Elbe in 1993. There is a single record from France (the Loire estuary).

A separate population lives in the Gulf of St Lawrence, where it represents the southernmost outpost of the species' natural distribution. Occasional strays have, however, been recorded farther south along the American east coast.

Belugas.

Pod of Belugas.

Because of its wide vocal repertoire, the Beluga is also called the sea canary. The sounds are not audible to the human ear without special equipment.

Reproduction: Males become sexually mature at an age of 8–9 years, while females are already sexually mature when 4–7 years old. Matings take place in the spring (April–May). The gestation period is between 14 and 15 months. Births occur at the end of July and the beginning of August. Lactation is estimated to last for up to 2 years. Twin births are known, but they are very rare.

Food: Fish, cephalopods and crustaceans.

Natural enemies: Killer Whales and Polar Bears.

Relationship with man: Some Belugas are killed in west Greenland and Canada. Accidental by-catches may occur in some places. The population in the Gulf of Saint Lawrence is seriously affected by pollution. Belugas are kept in captivity in several parts of the world, such as Chicago and Boston in the USA and Duisburg in Germany.

Similar species: In the main part of the range, adult Belugas cannot be confused with any other species. *Female Narwhals* could perhaps be mistaken for young Belugas, but the Narwhal's speckled pattern should normally lead to reliable identification. Old individuals of *Risso's Dolphin* can be almost entirely white and therefore perhaps be mistaken for Belugas, but Risso's Dolphin not only occurs farther south but also has a centrally placed dorsal fin.

When stranded, similar-sized species may at first glance be confused with Belugas in cases where decay has caused the outer skin to fall away and the stranded whale then appears white. The Beluga, however, will always be identifiable by its lack of a dorsal fin and by the particular shape of the flippers and tail flukes.

Narwhal.

Narwhal
Monodon monoceros

FRENCH Narwhal
GERMAN Narwal
SPANISH Narwhal

Description: Medium-sized, robust whale with rounded head, and comparatively short flippers which, in contrast to those of most other cetaceans, are flexible. On adults, the trailing edge of the tail flukes typically curves back beyond the level of the corners. Has no dorsal fin, but has a ridge in the middle of the back.
Size: Full-grown adults measure 4.2–4.7 m and weigh 800–1,600 kg; calves are 1.6 m long, with a weight of up to 80 kg.

Teeth: Typically, an anticlockwise-twisted tusk at the front in the left upper jaw (in most males and occasional females).
Coloration: The back and the flanks are mottled greyish-black and white, while the belly is white.
Variation: Males are longer and heavier than females, but their most striking feature is the tusk up to 3 m long on the left-hand side. A tusk can also grow from the right side, so that there are two tusks; the right-hand tusk is similarly anticlockwise-twisted and is typically a bit smaller than the left one. Females, too, can develop tusks, and big, very plump males without tusks also occur. Coloration changes from the juvenile's ash-grey through the typical mottled appearance to almost white in very old males.

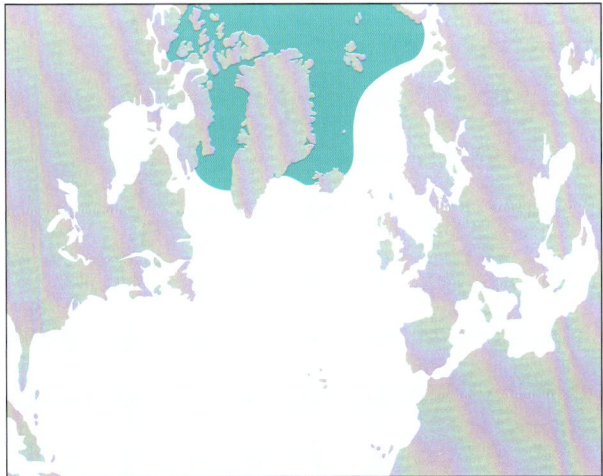

Distribution: The species' preferred habitat is the oceanic seas of the Arctic Ocean. In the North Atlantic, it has its main distribution in northernmost Baffin Bay, the northeast Greenland coast, Jan Mayen and Svalbard, while it is comparatively rare in the White Sea.

On rare occasions, odd individuals stray farther south. Narwhals have been seen along Iceland's north coast, off Scotland and, once, even as far south as the Øresund (between south Sweden and Denmark).

Behaviour: Groups typically consist of 2–10 animals. There are both male groups and mixed groups. Male groups often contain individuals of the same age, as some separation according to age occurs. The mixed groups consist of females with young of various ages. Movements are diffuse and nomadic.

The Narwhal is often seen spyhopping, tail-slapping and flipper-slapping, but it seldom leaps clear of the water. Dive times are between 7 and 20 minutes, and dive depths of several hundred metres are not uncommon. When surfacing, males reveal the tusk clearly. Males fight over the females by fencing with their tusks; these fights are sometimes so dramatic that the tusks splinter.

Reproduction: Sexual maturity is reached at an age of 11–13 years for males and 5–8 years for females. Matings take place in summer. The gestation period is 14–15 months, and lactation 20 months.

Food: Fish, cephalopods and shrimps.

Natural enemies: Attacks by Killer Whales on a Narwhal group have been recorded.

Relationship with man: Narwhals have for centuries been hunted for their tusks by Europeans and Eskimos. The Inuit people in Greenland and Canada today catch c. 1,000 animals annually.

Similar species: The only other species with a rounded head and of the same size is the *Beluga*. The Narwhal's tusk and mottled coloration render it normally unmistakable. Young Narwhals, however, can somewhat resemble Belugas, and very old male Narwhals can be almost white over the entire body.

Two Narwhals at the surface.

Characteristics of beaked whales.

Tail flukes without notch.

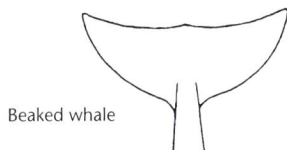

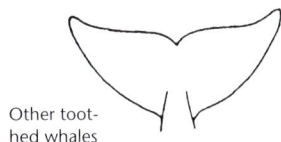

Beaked whale

Other toot-
hed whales

Throat grooves.

Very small flippers

Beaked whales

Beaked whales make up the second largest whale family (Ziphiidae), with six genera and at least 20 species. These are compact whales with a more or less well-defined beak, a dorsal fin set well to the rear on the back, and small flippers. Additional characteristics are the presence of throat grooves, and the fact that the tail flukes do not as a rule have a notch. Cephalopods represent the main prey of this family, as a result of which a significant reduction in the number of teeth has occurred. These generally erupt only in the males; in fortunate instances they may be visible in the field.

In the North Atlantic, six species are known to occur: four species of the beaked whale genus (*Mesoplodon*), the Northern Bottlenose Whale (*Hyperoodon ampullatus*) and Cuvier's Beaked Whale (*Ziphius cavirostris*).

The genus *Mesoplodon*, with at least 14 species, is the largest within the order of cetaceans. The true number is not yet established, as new species of this genus are continually being discovered.

There are only two teeth in this genus; they are in the lower jaw, either at the tip or set farther back. Also characteristic of this genus are the so-called 'flipper pouches', depressions in the body into which the flippers can be placed in order to reduce water resistance when swimming.

All the *Mesoplodon* species are oceanic, and are known primarily from strandings. This means that the distribution generally is poorly known.

Four species are treated here: *M. bidens*, *M. europaeus*, *M densirostris* and *M. mirus*. The possibility that new species may appear in the North Atlantic cannot be ruled out. In fact, this has already happened. In

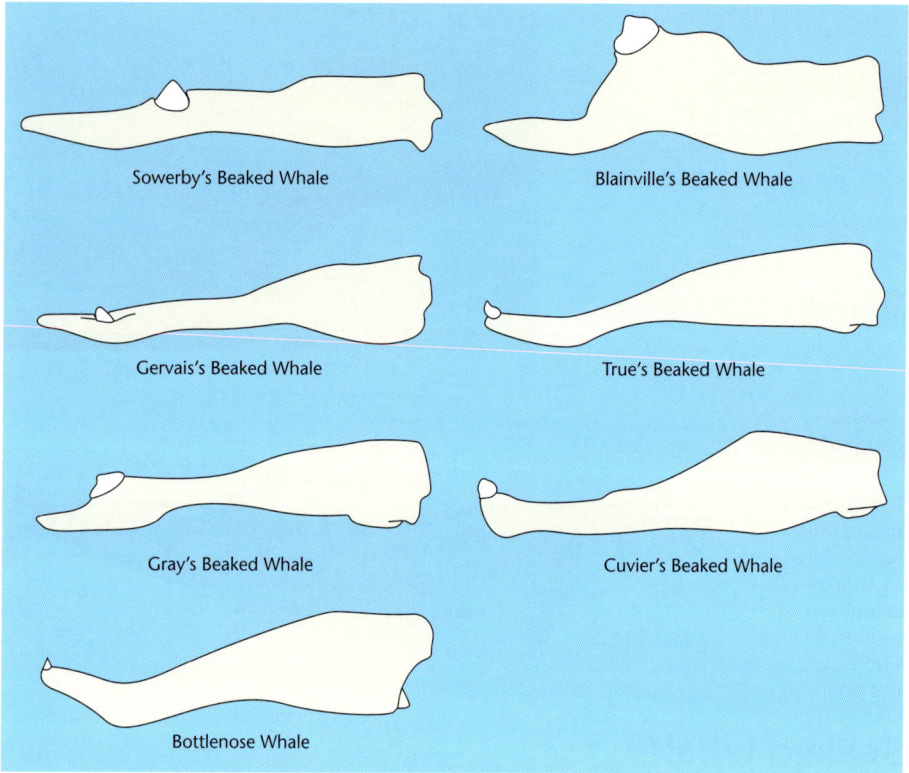

Lower jaw of the North Atlantic species of beaked whale.

1927, a single Gray's Beaked Whale (*Mesoplodon grayi*), a species which had until then been thought to occur only in the Southern Hemisphere, appeared on the Dutch coast. This individual has hitherto remained the only record of the species in Europe. While the possibility that a population of Gray's Beaked Whale exists in the North Atlantic cannot be excluded, this single record is considered to have involved a vagrant.

Reliable identification of beaked whales at species level can be made only on the basis of the teeth. Moreover, these can be seen externally only on adult males, so that females and juveniles can be identified only if opportunity allows the tooth to be worked out of the gum. In an ordinary field observation, it is to be hoped that females and juveniles are seen together with males. Indeterminate beaked whales are the rule rather than the exception.

Sowerby's Beaked Whale.

Sowerby's Beaked Whale
Mesoplodon bidens

FRENCH Baleine à bec de Sowerby
GERMAN Sowerby-Zweizahnwal
SPANISH Zifio de Sowerby

Description: Typical beaked whale. Corner of mouth is slightly upturned. Head has a more or less well-developed bulge at front and tapers into a beak. The flippers are rather short and are set in what are termed flipper pouches on the body sides. The throat grooves form an 'open V' with the open point directed forwards.
Size: Full-grown adults are 4.4–5.5 m long and 1,000–1,300 kg in weight; calves are c. 2.4 m long and weigh 170 kg.

Teeth: One tooth in each half of lower jaw about halfway between snout tip and mouth corner (visible only on adult males).
Coloration: The back is dark grey and merges into the paler flanks and the almost white belly. Curved scars from cookie-cutter sharks and white scratches from squid beaks are frequently visible. Typically, has a dark patch around the eye.
Variation: Males are presumably bigger than females. Teeth erupt only in adult males, which often show parallel, rail-like scars from mutual fights.

Behaviour: Very few data. Group size varies from 1 to 10 animals. Mixed groups occur, and presumably also male-only groups. Mother and calf probably stay together for over 12 months.

Distribution: The species' preferred habitat is the oceanic, temperate sea areas. It does not live in the North Sea, despite many erroneous statements to the contrary.

Has hitherto been recorded only in the North Atlantic, the central deep-water parts of which are believed to be the true core of distribution. Stray individuals have stranded along the entire northwest European coast. There is only a single record from the Mediterranean Sea. By far the majority of strandings are from the eastern North Atlantic, while on the American east coast the species is recorded most frequently north of Cape Cod Bay. Regularly seen in Cabot Strait, off the Canadian east coast. A stranding in the Gulf of Mexico, on the Florida coast, probably involved a stray.

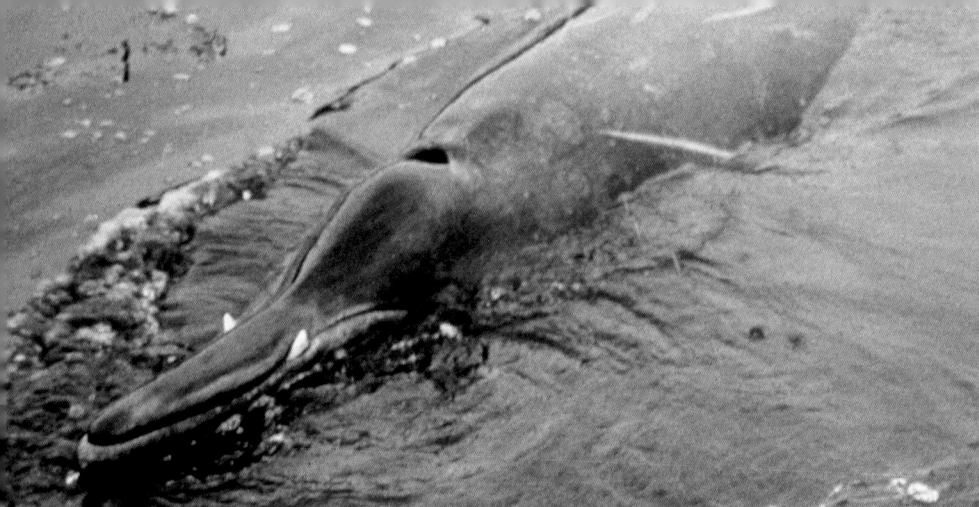

Sowerby's Beaked Whale.

Movements are not well documented, but possibly take place in conjunction with the advance of the ice. Along European and American coasts strandings are recorded throughout the year, but with a peak in the autumn months.

Regarded as being shy of vessels. The blow is normally not visible. Tail-slapping occurs. Dive times are normally 10–15 minutes, but can be up to 28 minutes with intervening periods at the surface of c. 1 minute. The beak is sometimes stuck out of the water before surfacing.

Reproduction: Almost nothing is known. Both sexes probably reach sexual maturity at a length of 4–5 m. No data are available on age at sexual maturity. Newborn calves have been recorded from June to September, and the smallest embryos from the summer period, which could indicate mating in spring. The length of gestation is not known, but is thought to be 10–12 months. A 3-m-long juvenile which was estimated to be c. 1 year old was still accompanied by its mother. On this basis, lactation can be estimated to last at least 12 months.

Food: Cephalopods and, to a lesser extent, fish.

Natural enemies: Potentially Killer Whales and larger sharks.

Relationship with man: A minor level of hunting has taken place in Newfoundland, but otherwise there has been no commercial exploitation of the species. Accidental by-catches in fishing equipment can occur. An attempt to return an individual to the wild after nursing it in captivity was unsuccessful.

Similar species: All other beaked whales in the region can potentially be confused with Sowerby's. Adult males of *Blainville's Beaked Whale*, however, are unmistakable owing to their large, powerful and visible teeth and the strongly arched mouth corners.

Gervais's Beaked Whale has straighter mouth corners, and the teeth are farther forward than on Sowerby's Beaked Whale. It seems to have a more southerly and easterly distribution than Sowerby's, but this criterion should be used only with great caution. *Gray's Beaked Whale* has been recorded once, in 1927, on the Dutch coast. As it has teeth in exactly the same position as Sowerby's, and as the possibility of the existence of a population in the North Atlantic cannot be excluded, every individual should be subjected to detailed scrutiny.

Blainville's Beaked Whale.

Blainville's Beaked Whale
Mesoplodon densirostris

FRENCH Baleine à bec de Blainville
GERMAN Blainville-Zweizahnwal
SPANISH Zifio de Blainville

Description: Typical beaked whale with strongly arched mouth.
Size: Full-grown adults are 4.7–5.3 m long and weigh 1,000–1,500 kg; calves are 2.3 m long and weigh c. 160 kg.
Teeth: There is one centrally placed tooth in each half of lower jaw (visible only on adult males).
Coloration: Blue-grey on the back and white on the belly. There may be yellow areas on the flippers and around the eyes and blowhole; this is due to the presence of diatoms on the skin. Scars from squid and cookie-cutter sharks are visible.
Variation: Males are bigger than females. The teeth erupt only in adult males; they can be extremely big and visible in the field when the animal is approached closely enough. Adult males are more scarred than females. The coloration of juveniles is possibly different, but good descriptions are not yet available.

Behaviour: Not particularly well known. Group sizes observed at the Canary Islands were of 2–9 animals, and groups consisted both of adult males and of females with young, but young males appeared to be absent. This could indicate that they leave the group and perhaps form juvenile groups, as with other toothed whales. The groups seem to be very tightly packed.

Distribution: Widely distributed in the world's warm-temperate, subtropical and tropical seas, in the Atlantic north to 40°N.

In Europe, it has been recorded on the coasts of Wales, the French Atlantic and Portugal, and on both the Atlantic and the Mediterranean coasts of Spain. It is also known from Madeira and the Canary Islands, from where there are year-round observations, especially from the sea area southwest of the islands of Gomera and Tenerife.

The species is known also from the American east coast, with strandings from Florida and the Bahamas in the south to Nova Scotia, Canada, in the north.

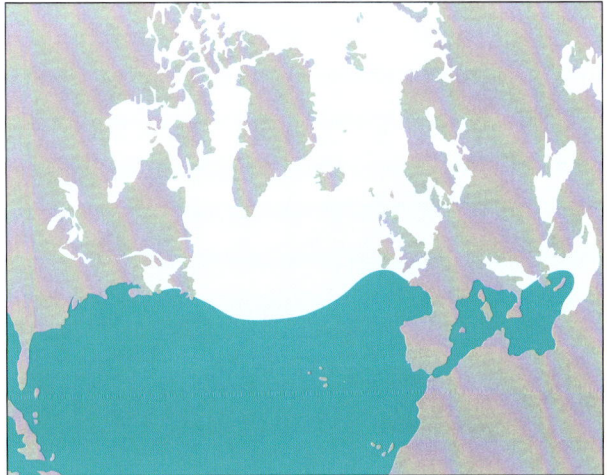

Blainville's Beaked Whale with calf.

Some individuals avoid ships, while others can be inquisitive and approach closely. Breaching has been observed, and head-showing and tail-slapping can also be seen. Regular dolphin leaps while swimming rapidly are also observed. Dive depths are at least 320 m. The animals dive in synchrony. Deep-dive times of 10 minutes are usual, but long deep-dive periods of over 45 minutes have also been recorded.

Reproduction: Both sexes are believed to be sexually mature at a length of 4.7 m. The youngest age at which females are mature is 9 years. There is no information on the mating season but, from the species' tropical–subtropical distribution, it may be assumed that matings take place throughout the year. There are no data on the duration of gestation and lactation, but these are probably similar to those of other beaked whales.

Food: Cephalopods.

Natural enemies: Bite marks on the tail of a stranded individual on the North Carolina coast indicate that Killer Whales or perhaps False Killer Whales may attack the species. Large shark species are also potential candidates.

Relationship with man: Possibly by-catches in fishing gear. There is no hunting of this species in the North Atlantic or elsewhere.

Similar species: Adult males, because of the particular tooth shape, cannot be confused with other species. In contrast, it can be extremely difficult to separate females from other beaked whale females. Best distinguished from the *Bottlenose Whale* and *Cuvier's Beaked Whale* by the head shape.

Blainville's Beaked Whale.

Gervais's Beaked Whale.

Gervais's Beaked Whale
Mesoplodon europaeus

FRENCH Baleine à bec de Gervais
GERMAN Gervais-Zweizahnwal
SPANISH Zifio de Gervais

Description: Typical beaked whale, with dorsal fin on posterior third of back and relatively short flippers. The jawline is almost straight.
Size: Full-grown adults are 4–5.2 m long and weigh 1,000–2,000 kg; calves are 1.6–2.2 m long and probably weigh c. 80 kg.
Teeth: A single tooth in each half of lower jaw, set in the anterior third of the jawline.
Coloration: Grey upperside with graduated transition to a white belly. A dark patch around the eye.

Variation: Females are possibly bigger than males. Maximum length: males 4.56 m, females 5.20 m. A white patch in front of the genital slit has so far been confirmed only for females.

Behaviour: Largely unknown. Group size is thought to be between 2 and 5 individuals. Information on group structure is lacking. Probably does not differ much from that of other beaked whales.

Reproduction: Very poorly documented. A male of 4.08 m was found to be sexually mature; the smallest mature female known was 4.2 m long. Duration of gestation and lactation is not known, but presumably

Distribution: The species' preferred habitat is the oceanic parts of the north and central Atlantic Ocean.
   Despite its scientific name (*europaeus*), it is very rare along European coasts, with only three records since its discovery in 1840 in the English Channel: in 1980 off the northwest Atlantic coast of Ireland, in 1986 off Portugal's west coast, and most recently in 1998 at the French Biscay coast. In addition, there are three records from the Canary Islands and two from the West African coast (Mauritania and Guinea-Bissau).
   Far more frequent on the western side of the North Atlantic. Along the USA's east coast this species is the commonest of the beaked whales. Strandings are known from Florida in the south to New York in the north.

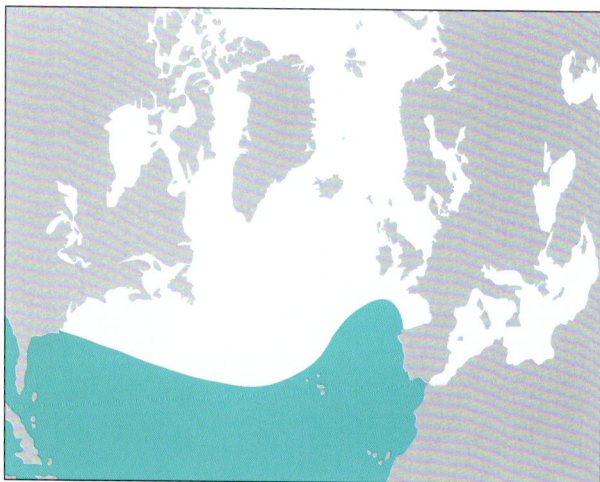

corresponds with that of similar species in the genus.

Food: Cephalopods.

Natural enemies: No information. Killer Whales and large sharks are, however, potential candidates.

Relationship with man: There is no commercial exploitation of this species, but accidental by-catches occur.

Similar species: Other beaked whale species. In the field, a reliable identification of the species is often not possible. Even when stranded, detailed inspection is often required in order to identify the species.
Sowerby's Beaked Whale has a more curved mouth and has the tooth set farther back than in Gervais's Beaked Whale. The latter apparently has a more southerly and westerly distribution than Sowerby's.
Blainville's Beaked Whale can be distinguished from Gervais's Beaked Whale by the strongly arched jawline. Adult males of Blainville's are unmistakable because of the visible teeth.

True's Beaked Whale
Mesoplodon mirus

FRENCH Baleine à bec de True
GERMAN Trues-Zweizahnwal
SPANISH Zifio de True

Description: Typical beaked whale, with rather narrow short flippers and a dorsal fin on posterior third of back. Head with slight bulge. Typically, has an eye patch.
Size: Full-grown adults are 4.9–5.3 m long and weigh 1,000–1,500 kg; calves measure c. 2.3 m in length and weigh c. 136 kg.
Teeth: There is one tooth at the tip of each half of lower jaw (erupts only in males).
Coloration: Grey upperside with graduated transition to a white belly. A dark eye patch may be present. Larger animals often show scars from squid and from fights with rivals.
Variation: The teeth erupt only in adult males, in which they may be visible at close range. Males are presumably bigger: maximum length for males is 5.33 m, for females 5.10 m.

Behaviour: Very little is known. Is never observed with certainty in the field, so that not even the group size is known. Probably does not, however, differ much from other beaked whales. Scratches and scars on adult males indicate that fights occur between them.

Reproduction: Very poorly documented, especially for the North Atlantic population. An almost fully developed embryo was recorded in a female which stranded on the northeast USA coast in March; in the Southern Hemisphere, a newborn calf has similarly been found in March. As the latter discovery was made in the austral autumn, matings and births presumably take place all year around—including in the North Atlantic.

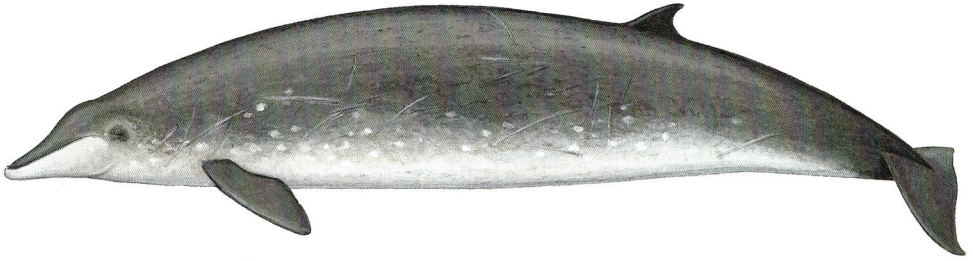

True's Beaked Whale.

Food: Cephalopods.

Natural enemies: No information. Killer Whales and large sharks are potential candidates.

Relationship with man: There is no commercial exploitation. Pelagic fisheries using floating trawl, purse seine or top nets in the deeper parts of the North Atlantic could potentially entrap individuals of this species.

Similar species: Other beaked whale species. In the field, reliable identification is often not possible. With stranded whales, a detailed examination is often required.

Distribution: The species' preferred habitat is deep temperate seas. Until 1959, when the first individual was found on South African coasts, it was assumed that the species occurred only in the North Atlantic.

In the eastern part of the North Atlantic, strandings have been reported from the southwest coast of Ireland and the Hebrides, the Spanish Biscay coast, and the Canary Islands and the Azores.

In the western part, the northernmost record is from Nova Scotia, Canada, and the southernmost from the Bahamas.

The relatively high number of records from the USA's east coast results from the close-knit reporting system. In Europe, observations of the species may be expected from the British and Irish coasts and in Bay of Biscay, as well as along West African coasts.

Northern Bottlenose Whale.

Northern Bottlenose Whale
Hyperoodon ampullatus

FRENCH Hypéroodon boréal
GERMAN Nördlicher Entenwal
SPANISH Calderon boreal, Ballenato boral

Description: Large beaked whale with high forehead and well-demarcated beak. The dorsal fin is set on the rear part of the back, and the flippers are relatively small. The blow of this species is normally visible; it is bushy in shape and 1–2 m tall.
Size: Full-grown adults are 5.8–9.8 m long and weigh 5,800–7,500 kg; calves are c. 3.5 m long at birth, while birth weight is not known but is estimated at c. 300 kg.
Teeth: There are typically 1–2 teeth in the tip of each jaw, the front one being big and the rear one small (normally erupt only in mature males).
Coloration: Blue-black to yellowish dark brown on head and flanks. The belly is paler. The head may show paler areas.
Variation: Males are bigger than females and develop a more bulbous head. There is wide variation with regard to the size, eruption and number of the teeth. In males the teeth first erupt at an age of 15–17 years, but even in older males they can be quite small and concealed in the gum. In some males, presumably rather old individuals, the teeth fall out.

Behaviour: Occurs singly or in groups of up to 35 animals. The structure is known

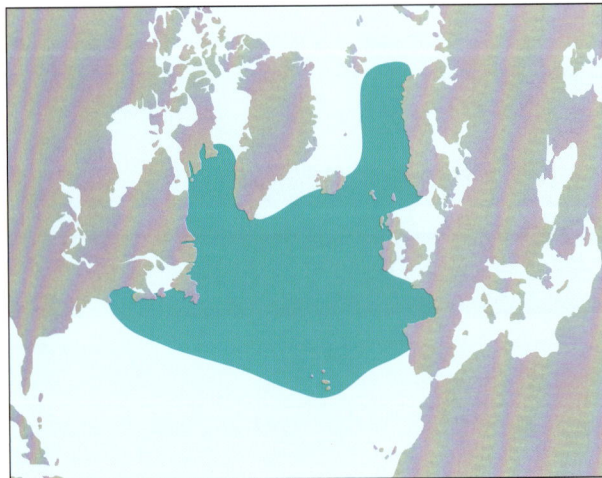

Distribution: The species' preferred habitat is the deep, offshore parts of the North Atlantic, where it has a patchy distribution and is locally very common.

In the eastern part of the North Atlantic, the species is known from the entire Norwegian coast, the waters between Iceland and Jan Mayen and southwest of Svalbard, the Faeroes and north Scotland. Periodically strays into the North Sea and the Baltic.

This whale occurs also at times in east Greenland, and is quite common in Davis Strait off northern Labrador, at the entrance to the Hudson Strait. Off the Nova Scotian coast it is found at The Gully. Also strays farther south along the American east coast.

from only few examples, but it seems that both mixed groups and male-only groups occur.

Migrates from summering waters in the north to wintering areas in warmer waters. At the Faeroes they appear in March, while they do not reach Jan Mayen until June, and the return movement begins as early as July.

Has been seen to leap clear of the water. In some places the animals exhibit great inquisitiveness and come right up to boats, whereas in others they are more shy. The greatest dive depth is 1,000 m or more. The whales can remain submerged for as long as 2 hours, but normally they are underwater for only between 14 and 70 minutes. Stays at the surface for up to 10 minutes at a time, where it blows every 30–40 seconds.

Reproduction: Males become sexually mature when they are 7.5 m long, females at 6.9 m. Age at maturity is 7 years for both sexes. Mating takes place in spring, and the gestation period is 12 months. Duration of lactation is at least 12 months. The female gives birth to one calf every second or third year.

Food: Fish, cephalopods and other invertebrates.

Natural enemies: Presumably only Killer Whales.

Relationship with man: The Bottlenose Whale was hunted commercially along Norwegian coasts, off Svalbard and Canada and also, to a lesser degree, in the Faeroes. From the end of the 1800s until about the 1930s, it is thought that perhaps as many as 80,000 animals were killed. After 1930 hunting was practised only on a small scale, with an estimated total catch of just over 5,000 animals. The species was given protection in 1977. Accidental by-catches in fishing gear occur.

Similar species: The Northern Bottlenose Whale can superficially resemble the two *pilot whales*, but a good look at the head with its distinct beak renders this species unmistakable. The position of the dorsal fin on the rear third of the back makes confusion possible with the *Minke Whale*, the dorsal fin of which can somewhat resemble that of the Northern Bottlenose Whale.

Northern Bottlenose Whales.

Cuvier's Beaked Whale.

Cuvier's Beaked Whale
Ziphius cavirostris

ALTERNATIVE ENGLISH NAME Goose-beaked Whale
FRENCH Ziphius
GERMAN Cuviers-Schnabeldelphin
SPANISH Zifio común, Ballenato de Cuvier

Description: Medium-sized beaked whale with poorly marked beak and distinct undershot jaw. The dorsal fin, which is comparatively small and sickle-shaped, is set on the hindmost third of the back. Unlike most beaked whales, it may have a weak marking or small notch on the trailing edge of the tail flukes; this, however, is never so distinct as in the other whale families.

Size: Full-grown adults are 5.1–6.9 m long and weigh 2,000–3,000 kg; calves are typically 2.2 m long, and weigh 250 kg.
Teeth: There is one tooth at the tip of each half of lower jaw (erupts only in adult males).
Coloration: Dark brown or grey on the back, where large areas of white are present. Scars from cookie-cutter sharks and, on males, from fights with other males.
Variation: Males are a little bigger than females. Old males may have an almost white head. The erupted teeth of males can be seen in the field at close range, and are more powerful than those of females.

Cuvier's Beaked Whale.

Behaviour: Occurs generally in groups of up to 25 animals, but solitary males are also seen. Group structure is not known in detail.

Normally avoids ships, but at times is inquisitive and comes close. Animals have been seen to leap right out of the water and land in a 'clumsy' fashion. When the species is travelling at great speed, the head may come into view. The whale can remain submerged for 20–40 minutes, and can probably dive to depths of several hundred metres.

Reproduction: Both sexes become sexually mature at a length of c. 5.5 m. Matings and births are thought to occur throughout the year. Gestation lasts 12 months, and lactation for at least the same amount of time. Gives birth on average to one calf every second or third year.

Food: Oceanic deep-sea cephalopods, fish and crustaceans.

Natural enemies: Presumably Killer Whales and large sharks.

Relationship with man: No commercial hunting of this species takes place, but accidental by-catches occur.

Similar species: Can be confused with other beaked whales, but at close range the head shape and, especially, the white areas make identification considerably easier.

Distribution: Occurs far offshore in the warm-temperate oceanic sea areas. In the eastern North Atlantic, the northern limit extends to the Scottish north coast and the Irish west coast. There are several records from British and Irish coasts, and the species has been recorded a number of times in Bay of Biscay. Outside this region, there are records from Sweden, North Sea records from the Belgian and British coasts, and one report from the Channel coast of France. The species is quite common in the deeper parts of the Mediterranean, and is seen around the Canary Islands. In Africa, it has so far been reported only from the Moroccan Atlantic coast, but may also be expected farther south along the West African coast.

Off the American east coast, the species appears especially in spring as far north as Newfoundland, and there have been several strandings on the USA coast (Massachusetts). The distribution follows the course of the Gulf Stream.

Sperm whales

The sperm whale family (Physeteridae) consists of both giants and dwarfs, and it has, not surprisingly, been questioned whether the two dwarf species (genus *Kogia*) and the big Sperm Whale really are closely related to each other. Some scientists emphasise the difference by placing the dwarf species in their own subfamily or even family. Here, all three species are, nevertheless, treated as belonging to the same family.

Size comparison of the three sperm whales.

Sperm Whale
Physeter macrocephalus

FRENCH Cachalot
GERMAN Pottwal
SPANISH Cachalote

Description: Big whale with compact build and with a large square head that accounts for up to a third of the body length. The position of the blowhole is quite unique among cetaceans, being sit-ed anteriorly on the left, instead of centrally on the top of the head. On the rear-most third of the back there is typically a large hump followed by a number of

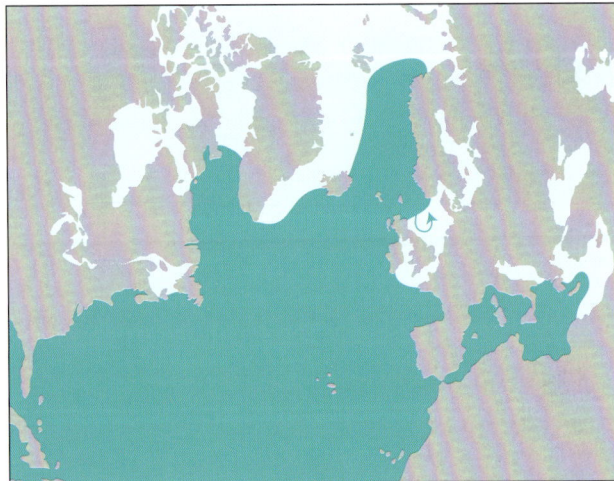

Distribution: The females normally live south of 45°N in the oceanic parts of the North Atlantic, but in recent years occasional females have stranded on the Irish west coast and groups of females have been observed in Bay of Biscay. This could indicate that the range is more flexible.

Big males occur on coasts of Greenland and Norway. Males migrate southwards in autumn. Strandings of Sperm Whales occur regularly in the North Sea, though this does not constitute a natural part of the species' range. It is predominantly, but not exclusively, young males that strand. The mass stranding on Rømø in 1997 included some animals over 50 years old.

Sperm Whale.

Group of Sperm Whales diving.

smaller humps. The flippers are almost rectangular. The skin has a characteristic wrinkled structure which has been likened to the texture of a plum stone.

Size: Full-grown males are 12–19 m long; females are somewhat smaller at 8–12 m. At birth the calf is c. 4 m long. Adults weigh 25–45 tonnes, while calves weigh up to 500 kg.

Teeth: There are 19–24 teeth in each half of lower jaw, which fit into cavities in the upper jaw. The gums of the upper jaw contain rudimentary teeth which sometimes protrude. The teeth do not erupt until the animal is 12 years old.

Coloration: Most of the body is uniformly dark brown to blue-black. White areas are present along the jaws and around the genital slit.

Variation: Males are significantly bigger and heavier than females. The large hump can be shaped almost like a dorsal fin, but may also be lacking altogether so that only the smaller humps are present.

Behaviour: Groups vary in size from 10 to 15 animals. Females form groups with juveniles of both sexes. Whereas females remain in these groups during puberty, males leave them at that time. Groups of young males then travel quite widely, i.e. farther northwards. Females defend the

The Danish island of Rømø was in two successive years the scene of mass strandings of Sperm Whales. Some of the animals from 1996 are shown here.

Sperm Whales' marguerite formation.

young against natural enemies, such as Killer Whales, by positioning them in the middle and adopting what is known as a marguerite formation: the mothers position themselves in a circle around the young, with the head turned inwards and the tail outwards facing the attackers.

Foraging is carried out by using powerful sound waves emitted from the large head; these are so powerful that they can paralyse or kill prey. When commencing a deep dive, the Sperm Whale shows its tail (*see* figure page 32). The blow is directed obliquely forwards. Sperm Whales can probably dive to depths of as much as 3 km and remain submerged for up to 2 hours.

Reproduction: Males reach maturity over several stages; even though they are in fact sexually mature at an age of 10 years, they do not become sexually active until they are over 19 years old and more than 13 m long. Females reach sexual maturity at 7–11 years and a length of c. 9 m. Young are born from July to November. Gestation lasts 14–16 months and lactation up to as long as 2 years (12–24 months). One calf is born every third to sixth year.

Food: Consists predominantly of pelagic squid, especially the species *Gonatus fabricii*, but also of large octopuses and, to a lesser extent, shoal fish.

Dwarf Sperm Whale.

Natural enemies: Reports exist of attacks by Killer Whales on groups of female Sperm Whales.

Relationship with man: Large-scale hunting of Sperm Whales has taken place since 1712 in the North Atlantic, from the American city of Nantucket, and this is thought to have led to a severe decline of the population as early as the end of the 1700s. This hunting, which was originally carried out with small boats and hand-held harpoons, survived until the late 1980s at Madeira and the Azores.

Until 1982 some Sperm Whales were also caught around Iceland and off the Spanish Biscay coast. Smaller-scale hunting has also taken place in Norway, the Faeroes and Ireland.

There are documented Sperm Whale strandings in the North Sea as far back as the 1500s, and the stranded animals were utilised locally for their oil in cooking etc. Throughout the 1800s, however, there were no strandings, which has been attributed to the heavy exploitation of the 1700s. The numerous strandings in the 1900s have been interpreted as a return of the population to its original size, but this idea has been refuted by others.

Similar species: The Sperm Whale cannot normally be confused with other species. The head shape and the skin texture are unmistakable. In addition, the diagonal blow is a good clue. Against the wind, however, the blow can be 'directed upwards'. The shape of the tail flukes is also quite characteristic.

Dwarf Sperm Whale
Kogia simus

FRENCH Cachalot nain
GERMAN Kleinpottwal
SPANISH Cachalote enaño

Description: Small, compact whale with centrally set dorsal fin. Has a shark-like appearance produced by the distinct overshot upper jaw, the projecting eyes, and the presence of 'bracket-like patches', which, because of their position on the body, are called 'pseudo-gills'. The blowhole is placed centrally on the top of the head.
Size: Full-grown adults are 2.1–2.7 m long and weigh 135–275 kg; at birth, the calf is c. 1 m long and weighs 40 kg.
Teeth: There are 7–13 teeth in each half of lower jaw, and 0–3 in each half of upper jaw.
Coloration: The back is dark grey and the underside white. Behind each eye there is a false gill.
Variation: There is no great difference between the sexes. No geographical differences have yet been described. The height and position of the dorsal fin vary somewhat, and in extreme cases there is overlap with the Pygmy Sperm Whale.

Behaviour: Observation of solitary animals or pairs is most frequent, but gatherings of up to 10 animals do occur. Group structure, however, is not known.

In situations of danger, these whales employ a 'squid tactic', in which they expel a cloud of excrement. They are shy, and seldom approach ships. The diving behaviour of the small sperm whales is highly characteristic: the animals simply allow themselves to drop, i.e. they sink passively without using flippers or flukes. Can probably dive to 300 m.

Reproduction: Both sexes become sexually mature at a length of 2.1–2.2 m. The age of sexual maturity is 3 years for males and 5–6 years for females. Matings and births may occur throughout the year. Gestation lasts 12 months and lactation also lasts c. 12 months. One calf is born each year.

Food: Cephalopods, fish and crustaceans.

Natural enemies: None reported, but Killer Whales and large shark species are potential candidates.

Relationship with man: There is no commercial exploitation of this species, but accidental by-catches in various types of fishing gear, chiefly bottom nets and floating nets, do occur. Live, stranded individuals have been taken into captivity to be cared for and later returned to the wild, but with little success.

Similar species: The only other species which it resembles is the *Pygmy Sperm Whale*, from which the Dwarf Sperm Whale was not taxonomically separated until 1966. The latter is smaller and its dorsal fin is in the centre of the back. On stranded animals, the height of the dorsal fin, which corresponds to more than 5% of the animal's length, is a good identification character. The Dwarf Sperm Whale floats deeper in the water than its close relative. The two species do, however, overlap in size, and the position and height of the dorsal fin are not totally reliable characteristics, either. On the other hand, it has been possible to differentiate the two species on the basis of blood-sample analysis.

The dorsal fin is not unlike that of the *Bottlenosed Dolphin*, but the Dwarf Sperm Whale has a completely different diving and swimming behaviour.

Distribution: The preferred habitat is warm oceanic sea areas, especially around shelf edges.

The Dwarf Pygmy Whale is rather common in parts of the Caribbean, and it is probably animals from here that reach the American east coast and Europe with the Gulf Stream.

Strandings are known from the Canary Islands and may be expected along the West African coast, although the latter area has not yet been sufficiently studied. There are three known strandings from the French Biscay coast, and a single record from each of the Italian and Portuguese coasts.

In the western part of the North Atlantic, there are many records in particular from Florida, but also regular sightings north to Cape Hatteras. The northernmost record is from Nova Scotia.

Pygmy Sperm Whale.

Pygmy Sperm Whale
Kogia breviceps

FRENCH Cachalot pygmée
GERMAN Zwergpottwal
SPANISH Cachalote pigmeo

Description: Small, compact whale with rectangular head and shark-like appearance (false gills).
Size: Full-grown adults are 2.7–3.4 m long and weigh 315–400 kg; newborn calves are 1.2 m long and weigh c. 44 kg.
Teeth: There are 10–16 teeth in each half of lower jaw and normally none in the upper jaw.

Coloration: Dark grey back and pale belly. 'Pseudo-gills' on rear head on each side.
Variation: There is probably only minimal difference in size between males and females. No geographical differences have been described. The shape of the dorsal fin varies somewhat, and in extreme cases it overlaps in both placement and height with that of the Dwarf Sperm Whale.

Behaviour: Group sizes of between 3 and 6 animals are known, but up to as many as 10 have been seen together. Group structure, however, is not known.

Distribution: Is probably more oceanic than its close relative. Observations are very rare, but strandings along North America's east coast and European coasts indicate a general distribution in the deeper, warmer parts of the North Atlantic. The North Sea is not normally part of the species' range, but a single stranding has been reported from Holland. In Bay of Biscay, a larger number of strandings are documented, but this does not mean that the species is concentrated in this area.

The species is recorded frequently on Florida's coast and is found north to Cape Cod.

Breaching is recorded. Uses the same 'squid tactic' as Dwarf Sperm Whale. Sometimes rests at the surface with the blowhole above the water and the tail drooping downwards, i.e. logging. When diving, it simply sinks in the water.

Reproduction: Females become sexually mature when they are 2.6–2.8 m long and 4 years old, while males reach sexual maturity when 2.7–3 m long and also at 4 years of age. Matings can occur throughout the year. Gestation lasts 11 months, and births can take place all year around. The lactation period is at least 11 months. A single calf is born each year.

Food: Predominantly cephalopods, but also shoal fish and crustaceans.

Natural enemies: Presumably Killer Whales and large sharks.

Relationship with man: The situation is the same as that for the Dwarf Sperm Whale.

Similar species: Very like the *Dwarf Sperm Whale*. The Pygmy Sperm Whale has a smaller dorsal fin (less than 5% of total length), which is set much farther back on the back, and it swims higher in the water than its close relative.

The two species overlap in size, and the position and height of the dorsal fin are not totally reliable characteristics. It has proved possible to distinguish between the two species by blood-sample analysis.

Pygmy Sperm Whale.

Baleen whales Mysticeti

The suborder of baleen whales (Mysticeti) contains four families, of which two occur today in the North Atlantic. These are the right whales (Balaenidae) and the rorquals (Balaenopteridae). A third family, the grey whales (Eschrichtiidae), was exterminated in the North Atlantic in the mid-1700s. The fourth family, the pygmy right whales (Neobalaenidae), is found only in the Southern Hemisphere.

The baleen whales are, in general, medium-sized or large animals. Common to them all is primarily, of course, the presence of baleen plates, which are modified, horny palatal folds. The baleen plates vary in shape, function and colour, and also in number, from species to species. The baleen whales lack teeth altogether.

Baleen whales have a double blowhole on the top of the head. As a rule, exhaled air, the blow or spout, is visible in the case of all these cetaceans.

Right whales

The right whales (Balaenidae) lack a dorsal fin and are completely smooth on the underside, i.e. they lack the throat grooves of the rorquals. They acquired the name 'right' because they behaved in the right way, the most appropriate way, for whalers to catch them. The head is big and accounts for up to a third of the animal's length. These are also quite plump whales. On some individuals, the maximum circumference (waist) measures 70% of the total length. The flippers are large and paddle-shaped. The baleen plates are very long, up to 5 m, and are used for skimming at the surface or in the upper layer of water. The right whales show the tail flukes on diving, and are characterised by having a V-shaped blow.

Two species occur in the North Atlantic: the arctic Bowhead Whale (*Balaena mysticetus*) and the temperate-water Northern Right Whale (*Eubalaena glacialis*).

Baleen from Bowhead Whale (left) and Fin Whale. The frayed side is directed inwards towards the centre of the mouth.

Bowhead Whale.

Bowhead Whale
Balaena mysticetus

ALTERNATIVE ENGLISH NAME
Greenland Right Whale
FRENCH Baleine de Groenland
GERMAN Grönlandwal
SPANISH Ballena groenlanda

Description: Big, robust baleen whale with large head. The head accounts for a third of the body length. The jawline is strongly arched. Has neither a dorsal fin nor throat grooves.
Size: Full-grown adults are 14–18 m long and can weigh 60–100 tonnes (second-heaviest whale); at birth, the calf is c. 4.5 m long, with an estimated weight of c. 1,000 kg.

Baleen: Has 250–350 black plates, 4–5 m long, hanging in two rows from the roof of the mouth.
Coloration: Black to blue-black over entire body, apart from white tail base and lower lip.
Variation: Females are a little bigger than males.

Behaviour: Groups of up to 6 animals are most frequent, but in very rare instances gatherings of up to 60 animals are seen. The latter cases, however, are the result of many animals aggregating simply because they are foraging in the same area, not because they are part of a permanent group.

Distribution: Formerly common throughout the Arctic Ocean, now extremely rare. Two discrete populations occur in the North Atlantic, one in the Davis Strait and one in the eastern North Atlantic in the area from Iceland to Jan Mayen and Svalbard. The northern limit of distribution is the edge of the ice, while the southern limit lies at 65°N.

Bowhead Whales are rarely seen in the North Atlantic. In recent years, however, single individuals and even small groups have been seen along Greenland's west coast; these belong to the Davis Strait population. The northeast Atlantic population is thought now to be almost extinct. Observations of individuals in this region could involve vagrants from other populations.

Slapping of the flippers and tail flukes is observed. Breaching is recorded. Swims very slowly. Typically, it remains for 1–3 minutes at the surface, where it blows 4–6 times, after which it dives. The blow is distinctly V-shaped. Shows the tail flukes on diving. Can remain submerged for up to 40 minutes. The tail-fluke display and the blow 'signature' are shown on pages 32 and 40, respectively.

Reproduction: Very little information. Females become sexually mature when they are 14–15 m long, males at 11–12 m. No method for reliable determination of age has yet been developed. Matings, probably involving sperm competition, presumably take place in January and February. Gestation is thought to last 12–14 months, and births occur in spring. The duration of lactation is estimated at 1 year, but there is some uncertainty over this. Females are believed to produce one calf every second or third year.

Food: Small crustaceans (copepods) and other invertebrates.

Natural enemies: Killer Whales.

Relationship with man: Hunting of Bowhead Whales began around Svalbard in the 1600s and lasted until the end of the 1800s, when the population was so depleted that hunting was no longer profitable. Hunting in the Davis Strait commenced in the early 1700s and continued until the early 1900s. Here, too, large catches led to almost total extermination of the stock.

The North Atlantic populations are now so low that they may be counted in just a few hundreds.

Similar species: The only other species having a V-shaped blow and which also shows the tail flukes on diving is the *Northern Right Whale*. That, however, has a more southerly distribution and, in contrast to the Bowhead Whale, has what are known as callosities on the snout and the lower lip.

The *Humpback Whale*'s blow, which is bushy or heart-shaped, can slightly resemble that of the right whales, but the clear differences between the two in habitat and behaviour will normally rule out any possibility of confusion.

Bowhead Whale.

Northern Right Whale.

Northern Right Whale
Eubalaena glacialis

FRENCH Baleine du nord
GERMAN Nordkaper
SPANISH Ballena del norte

Description: Big, robust baleen whale with large head. The head accounts for a third of the body length. The jawline is strongly arched. The flippers are large and fan-shaped. Has neither a dorsal fin nor throat grooves. Callosities are present both on the top of the head and in the centre of the lower lip.

Size: Full-grown adults are 14–18 m long and can weigh 80–100 tonnes; at birth, the calf is c. 4.5 m long, with an estimated weight of 1,000 kg.

Baleen: Has 200–270 brown plates, up to 3 m long, hanging in two rows from the roof of the mouth.

Coloration: Black to blue-black over entire body except for genital region, which is white.

Variation: Females are slightly bigger than males. No two individuals have exactly the same constellation of callosities, so that it is possible to identify animals individually from photographs (photo-identification).

Distribution: Formerly occurred both inshore and at sea in large parts of the North Atlantic. Today, a small relict population consisting of a few hundred individuals is found in the northwestern part. Summer distribution is concentrated in Bay of Fundy on the American east coast, while there are observations in the winter months from the Georgia coast and elsewhere. Migrations between these two regions take place in spring and autumn.

In the northeast part of the North Atlantic, it appears that only a few individuals remain. There are observations from the sea around the Faeroes and the Norwegian coast. These animals possibly constitute the remnants of a separate population, but it is more likely that they involve vagrants from the northwest Atlantic.

Behaviour: Groups of up to 6 animals are most frequent, but in very rare cases gatherings of up to 60 animals are seen.

Slapping of the flippers and tail flukes is observed. Breaching is recorded. Swims very slowly. Typically, it remains for 1–3 minutes at the surface, where it blows 4–6 times, before diving. The blow is bushy, V-shaped and 5 m tall. Shows the tail flukes on diving. Can remain submerged for up to 40 minutes. The tail-fluke display and the blow 'signature' are shown on pages 32 and 40, respectively.

Reproduction: Very little information. Females are sexually mature when 15 m long and 7–10 years old; males are mature at 14 m in length, but their age at sexual maturity is not known. Matings (and births) occur in winter off the Georgia and Florida coasts in the western North Atlantic. It is not known where the population in the eastern North Atlantic bred, but it may have been Bay of Biscay. Males have extremely large testicles, indicating sperm competition in association with mating. Females probably give birth to one calf every third or fourth year.

Food: Small crustaceans (copepods) and other invertebrates.

Natural enemies: Killer Whales.

Relationship with man: Hunting began perhaps as early as the year 1000 in Bay of Biscay, and led to the population in the northeast Atlantic being almost wiped out by the 1700s. The western North Atlantic population was exploited by Basque whalers from the mid-1500s, and in the 1600s American whalers arrived on the scene. The Northern Right Whale stocks were heavily hunted here, too, and from the mid-1700s the Americans switched instead to catching Bowhead Whales in the Davis Strait.

Today, the species is a popular target of whale-watching in Bay of Fundy and the Gulf of Maine, on America's east coast.

Similar species: The only other species with V-shaped blow and which also shows the tail flukes on diving is the *Bowhead Whale*. That, however, has a more northerly distribution and lacks the callosities shown by the Northern Right Whale.

Northern Right Whale.

Rorquals

The rorquals (Balaenopteridae), sometimes referred to as the fin whales, are characterised by the presence of throat grooves, which enable them to dilate the mouth enormously in association with foraging. The rorquals' method of feeding differs considerably from that of the right whales.

There are two genera in the family. The Humpback Whale is placed in *Megaptera*, which means 'big-wing' and refers to the large flippers of this stocky, robust whale. The other 5 species in the rorqual family belong to the genus *Balaenoptera*, which can be translated as 'winged whales', as they have long, narrow flippers, in contrast to the right whales' paddle-shaped flippers. These are slim, fast-moving cetaceans.

The overall build in the genus *Balaenoptera* is rather homogenous, and the individual species may be regarded as larger or smaller versions of one another—they have in fact been compared with Russian dolls. There is quite a big overlap in size between the individual species, and it can be difficult to identify them from a single characteristic, but by using a combination of several characteristics it is generally possible to distinguish them.

Under normal conditions the blow is expelled vertically upwards, and it is neither V-shaped nor forward-directed. Other good field characters are the position and shape of the dorsal fin; the bigger the whale, the farther back it is on the back. In

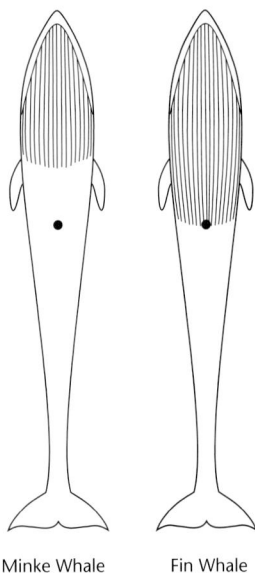

Minke Whale   Fin Whale

Variation in throat grooves of rorquals. The black spot indicates the position of the umbilicus.

the case of the smallest species, the Minke Whale, the fin emerges almost at the same time as the head when the animal surfaces.

With stranded individuals, the number and extent of the throat grooves are important features. It should, therefore, be noted how far back they extend and whether or not they reach the umbilicus (cf. figure above).

The blow characteristics of all species are shown on page 40.

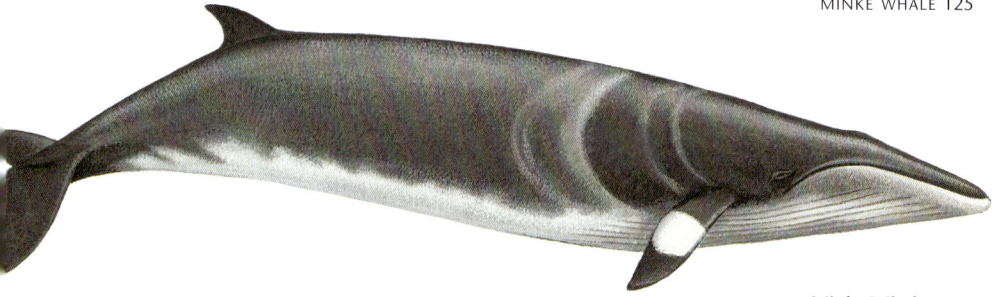

Minke Whale.

Minke Whale
Balaenoptera acutorostrata

FRENCH Balaenoptère acutorostrate
GERMAN Zwergwal
SPANISH Ballena picuda

Description: Sleek, medium-sized whale with very pointed head. The dorsal fin is falcate and comparatively tall. It is placed two-thirds of the way back from the tip of the animal's snout. The tail flukes are relatively broad (breadth equals a quarter of the animal's length). There are 30–70 throat grooves, which always terminate before the umbilicus (sometimes extending only slightly beyond the flippers).
Size: Full-grown adults are 6.75–10 m long and weigh 5–10 tonnes; at birth, the calf is c. 2.5 m long and weighs c. 350 kg.

Minke Whale.

Distribution: Oceanic and coastal throughout the North Atlantic, but rather rare in tropical waters. The distribution shifts northwards in summer and southwards in winter. Minke Whales can also be seen, however, in unexpected locations at unusual times.

Is observed along the entire Norwegian coast up to the White Sea. Also known from Iceland and the Faeroes. In autumn there are concentrations off the Scottish west coast, and in summer it is quite frequent in the northern North Sea and in the Skagerrak. Occasionally wanders into the Baltic Sea. Rare in the English Channel, but rather common in the Irish Sea and Bay of Biscay. May occur off the Atlantic coast of North Africa. Also found in the western Mediterranean.

Known from west Greenland in summer. Occurs along the American east coast south to Cape Hatteras, especially in April–November.

Minke Whale.

Baleen: Has 230–360 plates in each half of upper jaw; they are yellow, but have an increasing element of grey-brown towards the rear. They are from 12 cm (front) to a maximum of 20 cm (rear) long.

Coloration: Black back with variable whitish or bleached areas. Sometimes has an angular patch behind the head. The flippers have a white band on the outer part.

Variation: Females are slightly bigger than males. The pattern on the back can be used to identify individuals. On some, there is a rather sharp marking between the pale and the dark areas on the back, behind the head. No geographical differences in colour pattern have yet been described for the North Atlantic.

Behaviour: Normally forms small groups of up to 3 animals. Larger gatherings associated with foraging are known. It is thought that geographical separation by sex and age occurs.

The Minke Whale is reported to have migrated up to 9,000 km, but populations which are largely resident also exist.

Swims comparatively quickly. Spyhopping and breaching are reported for the species. Some individuals are rather inquisitive and are attracted by smaller boats (there are instances of individuals having accompanied a whale-watching boat for more than half an hour); others are more shy. The tail flukes are not shown when diving. The whale blows 5–8 times at intervals of 3–8 minutes, but the blow is very weak; it is seen at the same time as the dorsal fin appears. Can remain submerged for up to 20 minutes. Dive depth is not known for certain, but, as this species feeds on surface animals, it probably does not normally exceed 50 m.

Reproduction: Females and males become sexually mature at c. 6–8 years of age, when they are, respectively, 6.75 m and 7.32 m long. Matings (and births)

take place from winter to early spring. Gestation lasts 12 months, and lactation 3–6 months. Gives birth to a single calf every 1 or 2 years. Twin births are not known for this species.

Food: Shoal fish such as herring and capelin, and small crustaceans.

Natural enemies: Presumably Killer Whales.

Relationship with man: For many years, the hunting of Minke Whales was pursued along the Norwegian coast, and off Greenland, Iceland, Canada and other places. With the exception of Greenland, this whaling ceased in 1985, when a 10-year total protection of all commercially exploited whale stocks was agreed upon. Since 1994, hunting of this species has again been carried out by the Norwegians; this involves 200–400 animals being killed each year along the Norwegian coast and in the northern North Sea. Greenland whalers have an annual quota of 150 animals.

Minke Whales can also become entangled in fishing gear, and thereby be killed unintentionally.

Similar species: At longer distances the Minke Whale can be confused with other rorquals, especially the *Sei Whale* and *Bryde's Whale*, which are closest to it in size. The weak blow and the surfacing sequence, where blow and dorsal fin appear together, should, however, be good characteristics. The Minke Whale arches the tail stock far more than do the two other rorqual species.

At close range, the pointed head and the white bands on the flippers are unmistakable. The *Northern Bottlenose Whale's* dorsal fin is somewhat similar, both in size and in placement, to that of the Minke Whale; its head shape, however, is very different.

Bryde's Whale
Balaenoptera edeni

FRENCH Rorqual de Bryde
GERMAN Brydewal
SPANISH Rorcual tropical

Description: Medium-sized to slender whale, typically with three ridges on top of the head. The dorsal fin is distinctly falcate, almost forming a right-angle. It is comparatively tall (up to 50 cm) and is set on the posterior third of the back. The underside of the tail flukes may be a dirty white. There are 40–70 throat grooves, which extend from the snout typically well beyond the umbilicus.

Size: Full-grown adults measure 11.5–14.5 m and weigh 12–20 tonnes; newborn calves are 3.4–4 m long and weigh c. 900 kg.
Baleen: Has 250–370 ash-grey plates c. 40 cm long in each half of upper jaw.
Coloration: Dark grey on back, but with white or paler areas. Underside is white.
Variation: Females are slightly larger than males. In some regions, coastal and oceanic forms of this species have been described. The coastal form is smaller and perhaps occupies the same niche as the Minke Whale in the southern part of the North Atlantic.

Distribution: Occurrence in the North Atlantic is not particularly well known, but the species' general distribution lies south of 40°N. It inhabits the warmer waters, often close inshore, where there are upwellings and therefore concentrations of its prey.

In the eastern North Atlantic, the range of Bryde's Whale probably extends to the Canary Islands; it is possibly found as far north as Bay of Biscay.

In the western North Atlantic, the distribution reaches north to Cape Hatteras on the east USA coast.

Bryde's Whale.

Behaviour: Typically in ones and twos, but loose groups of up to 30 animals are seen.

The migrations are very diffuse, but in some places there seems to be a seasonal pattern resembling that of other rorquals, even though the movements are over far shorter distances. Other populations can probably be regarded as more or less resident.

Bryde's Whales at times come close up to boats and smaller ships. Spyhopping is observed. When foraging, they roll on to their side and sometimes carry out 'half-hearted' breaching. The tail flukes are normally not raised when diving. The whale blows on average 4–7 times before diving: the blow is 3 m tall and elliptical. Dives to greater depths than the Sei Whale, as its prey is found in deeper water. Typically remains submerged for 2–8 minutes. Surfaces at a steep angle.

Reproduction: Males become sexually mature at a length of c. 12 m, females at 12.5 m. Gestation lasts 1 year, and the ensuing lactation period less than 12 months. Females give birth to one calf on average every other year.

Food: Fish and krill.

Natural enemies: Presumably Killer Whales.

Relationship with man: Has been commercially exploited on only a small scale. The true extent of this is difficult to assess, as information on catches of the Bryde's Whale is mixed with catch statistics for Sei Whales. The Bryde's Whale was not distinguished as a separate species from the latter until as recently as 1912.

Similar species: Most similar to the *Sei Whale*, which is, however, bigger and has only one central ridge on the head, whereas the Bryde's Whale has three. Stranded animals can be identified by the colour and fineness of the baleen: Bryde's Whale's baleen plates are 'stiff brushes' in comparison with the Sei Whale's fine fringes.

At longer distances, confusion can arise with the *Minke Whale*. The Bryde's Whale, however, never has the latter's white marks on the flippers and behind the head.

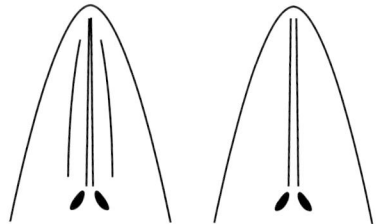

Bryde's Whale (left) has three ridges on the upperside of the head, whereas other rorquals normally have only one.

Sei Whale
Balaenoptera borealis

FRENCH Rorqual de Rudolphi
GERMAN Seiwal
SPANISH Rorcual del norte

Description: Large, slender whale with an erect, comparatively tall dorsal fin set on the posterior third of the back. The tail stock is relatively thick. Has 32–62 throat grooves which typically end between the flippers, before the umbilicus.

Size: Full-grown adults are 12–18 m long and weigh 20–30 tonnes; calves measure 4.5 m and weigh c. 725 kg.
Baleen: Has 300–410 greyish-black, finely fringed plates in each half of upper jaw.
Coloration: Predominantly blue-grey to dark grey on back and flanks and white on belly. The flippers are black on both inner and outer parts. May have circular white scars on the flanks from cookie-cutter sharks and lampreys.

Dorsal fins of Fin Whale (upper) and Sei Whale.

are normally of between 5 and 20 minutes. The maximum dive depth is thought to be a few hundred metres.

In the eastern North Atlantic, migrations probably take place in spring and autumn between feeding areas in the north and breeding grounds in the south. Where the latter are situated, and their limits, are not known. Similar migrations appear to occur in the western parts of the North Atlantic, but, again, the location of the wintering (breeding) grounds of the Sei Whales recorded in summer on the American east coast remains to be discovered.

It is interesting to note that this species first became known to science when an individual was found stranded at Grömitz, on the Baltic coast of Holstein, in 1819.

Sei Whale.

Variation: Females are a little bigger than males. Some Sei Whales have a more falcate dorsal fin, with the tip of the fin curved backwards.

Behaviour: Groups tend to be small, of 2–5 animals, although larger gatherings of up to 30 may be seen in the feeding areas. The Sei Whale can travel almost as fast as the Fin Whale. It seldom breaches. It normally blows every 20–30 seconds when at the surface for 1–4 minutes. When diving, it does not show the tail flukes. Dive times

Reproduction: Both sexes become sexually mature at an age of 10 years, when males are 13 m long and females 13.5 m. Matings (and births) take place in winter. Gestation lasts 10–12 months, and lactation 7 months. Females give birth to one calf every second or third year.

Food: The prey consists almost exclusively of copepods and bioluminescent krill. More rarely, shoal fish such as capelin

Distribution: The Sei Whale occurs throughout the North Atlantic, but keeps to the temperate and subtropical, oceanic parts. It avoids the colder arctic waters. In the northeast Atlantic, it is recorded in summer southwest of Iceland, at the Faeroes, Norway and Scotland. The species is very rare in the North Sea and the Baltic Sea. The Sei Whale is known from the Spanish Atlantic coast, but it is unclear whether it occurs in the westernmost Mediterranean. Its presence on the West African coast is not well documented, and is complicated by the occurrence of Bryde's Whale in the same area.

On the American east coast, Sei Whales are recorded at Cape Cod and Georges Bank in summer.

Sei Whale.

have also been found in the stomachs of North Atlantic Sei Whales.

Natural enemies: Presumably Killer Whales and larger sharks.

Relationship with man: There has been some commercial hunting of Sei Whales in Iceland, Norway, Ireland, Scotland and Spain. This ceased in 1986, but not until 1990 in Iceland. Is possibly subject to accidental by-catches in various kinds of fishing gear.

Similar species: Can be confused with the *Bryde's Whale* and possibly with the *Minke Whale*, which are closest to it in size. Unlike Bryde's Whale, the Sei Whale has only one central ridge on the top of the head. The Minke Whale has obvious white markings on the flippers and behind the head. The surfacing sequences are also rather different: with the Minke Whale, blow and dorsal fin come into view almost simultaneously, whereas with the Sei Whale there is a brief interval between the blow being seen and the appearance of the dorsal fin. The Sei Whale's dorsal fin is relatively farther forward than that of other rorquals. It can best be distinguished from the *Fin Whale* by the shape and angle of the dorsal fin: on Sei Whale the front edge of the fin creates a right-angle, whereas on Fin Whale it forms an angle of c. 135°.

Fin Whale.

Fin Whale
Balaenoptera physalus

FRENCH Rorqual commun
GERMAN Finnwal
SPANISH Rorcual común

Description: Large, slender whale with a comparatively low dorsal fin on the posterior third of the back. The front edge of the fin forms an angle of c. 135° with the back. Has 50–100 throat grooves, which reach the umbilicus.

Size: This is the world's second-longest animal species. In the North Atlantic, adults are 17–24 m long and weigh 30–80 tonnes; calves are c. 6.5 m long and weigh c. 2 tonnes.

Baleen: Has 260–480 plates in each side of the upper jaw. In the anterior third to half of the right side they are yellowish with dark stripes; in the rest of the right side and the entire left side the baleen plates are dark.

Coloration: The back is blackish-grey, and the underside white except for the head, which is asymmetrical in colour: the left side is dark, while the right is white. The asymmetrical head coloration is, as already noted, reflected in the baleen plates. An angular white mark is present behind the head.

Variation: The sexes differ: females are marginally bigger than males. On some individuals, the head coloration is the reverse of that described above. No geographical differences in coloration are known.

Distribution: The Fin Whale lives in the open sea, but also at times in inshore areas with deep fjords. Its migrations presumably follow broadly the pattern of other rorquals, but are more diffuse or more flexible.

In the eastern North Atlantic, the species occurs along Norway's coasts up to Svalbard and around Iceland. Sometimes appears in the North Sea, Skagerrak, Kattegat and the Baltic. In the Mediterranean, there is a population in the Ligurian Sea between Corsica and the Italian and French mainland.

It is well known along Greenland and Labrador coasts and occurs also on the American east coast south to Cape Hatteras. It appears relatively frequently in the Gulf of St Lawrence.

Behaviour: Groups are as a rule small and consist of 3–7 animals. Up to 100 animals may, however, assemble in the feeding areas.

Breaching and flipper-slapping are well known for this species. Fin Whales apparently do not react to the presence of ships, but merely continue to behave as they were. With a top speed of just under 40km/h, the Fin Whale is one of the fastest of all rorquals.

It blows typically 2–5 times at intervals of 10–20 seconds before diving. The blow is distinct and 4–6 m tall. The Fin Whale only very rarely shows its tail flukes when diving; when it does, it is usually young animals that are involved. Dive times are generally of 5–10 minutes, but they can be longer. Dive depth is at least 230 m.

Reproduction: Becomes sexually mature at an age of 6–12 years and a length of 18.3 m and 17.7 m for females and males, respectively. Matings normally take place from October to January. As they are not confined to the warmer sea areas, they can occur all year around, as has been documented from stranded newborn calves on the British North Sea coast. Gestation lasts 10–12 months, and births take place in the same period as matings. Lactation lasts 6–7 months. The female gives birth to one calf every second or third year.

Back of Fin Whale.

Fin Whale lying on its side and showing flipper.

Food: The Fin Whale's diet is very catholic. It includes fish, cephalopods and crustaceans. Foraging is not restricted exclusively to the colder sea areas, but occurs also in warmer waters where upwellings are found. The asymmetrical coloration is important for foraging, as the whale swims with the right side facing downwards.

Natural enemies: Killer Whales.

Relationship with man: Fin Whale stocks in the North Atlantic were among the first to be exploited following the invention of the harpoon gun in the 1860s and the adoption of the faster steamships for whaling. Hunting of Fin Whales was carried out in the eastern North Atlantic from Norwegian coasts and at Svalbard, Iceland, the Faeroes, Scotland and Spain, and from west Greenland and Newfoundland in the western North Atlantic. In Greenland, a small number of Fin Whales are still killed by the local population for its own use.

Similar species: The Fin Whale's asymmetrical coloration and characteristic dorsal fin will normally lead to a reliable identification at close range. The *Sei Whale's* dorsal fin is comparatively taller and set at right-angles on the back. The *Blue Whale* usually has a very small dorsal fin and a blue-grey coloration.

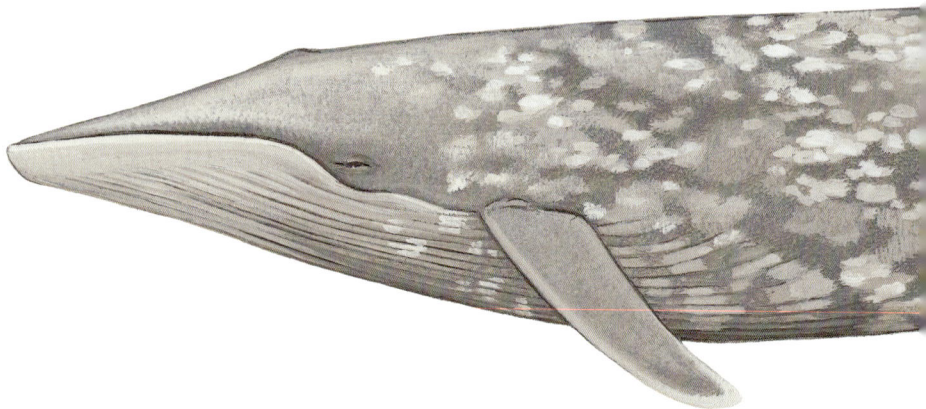

Blue Whale
Balaenoptera musculus

FRENCH Rorqual bleu
GERMAN Blauwal
SPANISH Ballena azul

Description: Gigantic but streamlined whale. The dorsal fin is often quite small and is set on the rearmost quarter of the back. The tail flukes from tip to tip are very broad, spanning a distance equivalent to a

Blue Whale.

quarter of the animal's total length. Has 55–88 throat grooves.

Size: The Blue Whale does not grow so big in the North Atlantic as do its conspecifics in the northern Pacific or in the Southern Hemisphere. Even so, the maximum length is enormous, at 28 m. Adults are 23–28 m long and weigh c. 125 tonnes; calves are 7 m long and weigh 2.5 tonnes.

Distribution: The Blue Whale population in the North Atlantic has been greatly depleted, and today there are perhaps fewer than 500 individuals left. These are concentrated particularly in the western sector. Blue Whales are seen with some regularity off Greenland's west coast. Along west European coasts, strandings of this species have not been recorded since 1936. In recent years, however, individuals have again been observed at the Azores and in Bay of Biscay.

Tail flukes of Blue Whale.

Baleen: Has 270–395 black plates up to 1 m long in each side of the upper jaw.

Coloration: The body is grey-blue all over, but the throat grooves can have a pale blue-grey colour.

Variation: The sexes differ: females are slightly bigger than males. As a matter of interest, it is worth mentioning that crosses between Blue Whale and Fin Whale have been found on several occasions, and that these hybrids of female sex have in fact produced viable offspring after mating (or back-crossing) with a Blue Whale.

Behaviour: Occurs solitarily or in twos. Many more animals sometimes gather together in the feeding areas.

Certain individuals are approachable, while others avoid ships. Only young animals exhibit aerial behaviour with breaching. Full-grown adults make do with tail- and flipper-slapping. Can accelerate to speeds of up to 30 km/h. Dive times, diving sequences and dive depths depend on the whale's activity. If the whale is relaxed, it blows every 10 to 20 seconds for 2–6 minutes before diving. The blow is tall, and the tail is raised when diving. Blue Whales remain submerged for between 5 and 20 minutes, and the maximum dive depth is 150 m. On surfacing, the dorsal fin emerges after the head has broken the surface.

Reproduction: Both sexes become sexually mature at an age of c. 8 years, when females and males are respectively 23 m and 21 m long. Matings take place in winter. Gestation lasts 10–12 months, and births take place in winter. They are followed by a lactation period of 7 months. The calf puts on 91 kg in weight per day, or 3.75 kg per hour!

Food: Consists exclusively of bioluminescent krill.

Natural enemies: Killer Whales. Large shark species may also represent a danger to young.

Relationship with man: Subjected to very intensive whaling from the 1860s to the middle of the 1900s, with the participation of Norway, Iceland, Scotland and Spain, but the species' importance in relation to the total whale catch diminished quite quickly.

Today, the Blue Whale is the object of whale-watching in the Gulf of Maine, where there is a photo-catalogue of several hundred individuals.

Similar species: The Blue Whale's size and coloration, together with the tiny dorsal fin, make it unlikely to be confused with any other species. The only other whale which comes anywhere near it in size is the *Fin Whale.* That, however, has a dorsal fin which is bigger and is quite different in shape.

Humpback Whale.

Humpback Whale
Megaptera novaeangliae

FRENCH Baleine à bosse, Mégaptère
GERMAN Buckelwal
SPANISH Rorcual jorobado

Description: Medium-sized whale with robust build. The flippers are characteristically one-fifth the length of the body. Has sensory tubercles on the upper lip, a stubby dorsal fin, and 14–35 throat grooves.
Size: Adults are 11.5–15 m long and weigh 25–30 tonnes; newborn calves are 4.5 m long and can weigh 1–2 tonnes.
Baleen: The plates are up to 1 m long and black to olive in colour. Has between 270 and 400 in each side of the upper jaw.
Coloration: The back is blue-black or dark grey. Underside can be white to a varying extent. The flippers are white. The underside of the tail flukes can vary from all white to all black.

Distribution: The Humpback Whale can be seen both in oceanic waters and along coasts; it can travel easily in complex coastal waters, even river mouths and the lower reaches of rivers.

The species occurs throughout the North Atlantic, but at differing times. In summer it lives in the feeding areas in New England, Greenland, Iceland and off the north coast of Norway, while in winter it is found in the breeding areas off Haiti and Puerto Rico. Single Humpback Whales have again been seen at the former breeding area of Cape Verde.

In the eastern North Atlantic, Humpback Whales are reported in the White Sea and down the Norwegian coast. In the North Sea and the Baltic they are relatively rare, but are still encountered. In the latter sea, there are reports dating back to the 1700s. In the western North Atlantic, the species occurs rather frequently in the Gulf of Maine and adjoining seas, as well as off west Greenland.

Humpback Whale breaching.

Tail flukes of Humpback Whale.

Variation: The sexes differ: females are bigger than males. The flippers and the underside of the tail flukes vary, allowing individuals to be identified by using photographs (photo-identification). The North Atlantic's Humpback Whale population is one of the best-studied mammals with regard to genetic variation.

Behaviour: Humpback Whales normally occur in small pods of 2–15 animals; larger numbers congregate in productive feeding areas. They migrate long distances between mating/breeding areas and feeding areas, though younger, juvenile individuals may stay behind in the feeding areas. Even though animals from different regions mingle in the breeding places, the females almost always return to the same feeding area; there are few exceptions to this rule.

Humpback Whales are rather sluggish, generally swimming at a speed of only 8 km/h at the surface. During mating, males may give powerful slaps of the tail or flippers. Breaching can sometimes carry them clear of the water. The record for breaching is 60 leaps in succession.

Humpback Whales have developed several spectacular feeding techniques. Most familiar is the so-called bubble-net, which entraps fish shoals. Rapid lunges (lunge-feeding) at fish shoals are also seen. The males sing to attract females in association with mating competition. The song is identical for all males in the same year, but it changes very slightly from year to year. It has been shown that there is also another song, which occurs outside the breeding season.

Blows 4–10 times at intervals of 20–30 seconds, before diving. The blow is bushy and up to 3 m tall. Before deep dives the tail stock is arched, so that the tail flukes come into view. It dives to depths of several hundred metres and can remain there for 3–7 minutes.

Reproduction: Humpback Whales become sexually mature at an age of 4–6 years; males are then 11.5 m long and females 12 m. Matings (and births) take place in well-defined shallow-water breeding areas off Puerto Rico and Haiti. Gestation lasts 11 months and is followed by a lactation period of c. 1 year. Females produce one calf every second or third year.

Food: Fish, aggregations of cephalopods and bioluminescent krill.

Natural enemies: Killer Whales.

Relationship with man: Hunting of Humpback Whales stopped altogether in 1996, with protection of all stocks. The populations in the North Atlantic, however, were already greatly depleted in the 1800s. In more recent times, accidental by-catches and collisions with ships have become the greatest risks.

The Humpback Whale is a popular animal for whale tourism. Whale-watching for Humpbacks is carried out in Greenland, Iceland, Norway and Newfoundland, as well as in the Bay of Fundy.

Similar species: Owing to its peculiar appearance, the Humpback Whale is unlikely to be confused with other cetaceans. Individuals with a less well-developed hump beneath the fin can perhaps be confused with the *Fin Whale* or the *Minke Whale*, but the white flippers are unmistakable. If the Humpback shows its tail flukes, it can be distinguished from other large whales by the flukes' jagged trailing edge.

Humpback Whale spyhopping.

SEALS AND WALRUSES

Diversity and classification

The seals have recently been included in the order Carnivora, but the traditional classification, which treats them as an independent mammalian order (Pinnipedia), is followed here.

There are 34 species, divided into three families: eared seals, walruses, and true seals. The true seals are further subdivided into two subfamilies: Phocinae, which is restricted to the Northern Hemisphere, and Monachinae, which is found chiefly in the Southern Hemisphere but is also represented in the Northern Hemisphere by the monk seals.

In the North Atlantic there are eight pinnipeds: the Walrus and seven true seals.

Origins and adaptations

All seals originate from terrestrial carnivores. It was previously thought that the seal order was an 'unnatural', polyphyletic, unit, because it was believed to contain two unrelated groups: on the one hand, eared seals (sea-lions and fur seals) and walruses, which were said to have evolved from bear-like carnivores, and, on the other, the true seals, which were regarded as most closely related to the mustelid family among the terrestrial carnivores. New research has, however, shown that all three seal families originate from the same group of land-dwelling carnivores, and that true seals and walruses are in fact each other's closest relatives within the order.

Comparative plate of seals.

Grey Seal, page 160

Common Seal, page 153

Bearded Seal, page 162

Ringed Seal, page 156

Hooded Seal, page 165

Harp Seal, page 158

Mediterranean Monk Seal, page 167

Unlike the cetaceans, the true seals have retained many of the characteristics of the land mammals. They have a pelt and four limbs. In terms of their body shape, the individual seal families exhibit different degrees of adaptation to a life in water. The eared seals, as their name suggests, still have external ears, whereas these are completely absent on the true seals. Eared seals and walruses can support themselves on their hindflippers when moving on land. In the true seals, the hindlimbs do not support the body on land, so the animals use only the forelimbs when advancing over sandbanks etc.

Seals are equipped with an insulating layer of blubber. In the true seals, the blubber is the only really effective insulation, whereas the fur seals also have a layer of hair for this purpose.

Despite being adapted for an aquatic life, none of the seal families is totally independent of land. All give birth to their young on land (or ice), and also spend much time on land when moulting.

Reproduction

For most species, matings and births take place within a comparatively narrow time interval, and 11–12 months elapse between mating and birth. The fertilised egg does not attach itself to the placenta for a number of months. This is known as delayed implantation, and occurs also in some other mammal species. Embryonic development itself lasts between 6 and 7 months. The fact that the pups are large from birth, and put on weight rapidly, ensures a high survival rate of young.

The majority of seals have two nipples, but a few exceptions exist. Bearded Seal and monk seals have four.

Seal skull showing typical dentition.

Dentition

The teeth are divided into *front teeth* (incisors, shortened to I), *eye teeth* (canines, C) and *cheek teeth* (post-canines, sometimes divided into premolars and molars, and abbreviated to PC, PM and M), and there is both a set of milk teeth and one of permanent teeth. A dental formula describes the number and placement of the teeth: e.g. 'I3/3, C1/1, PC5/5' indicates three front teeth (I) in both upper and lower jaws (3/3), one eye tooth (C) in both upper and lower jaws (1/1) and five cheek teeth (PC) in both upper and lower jaws (5/5).

Senses

The seals utilise sense of smell in the same way as do land mammals, which includes recognition of young and presumably other forms of communication. This is because seals, like land mammals but unlike cetaceans, possess sebaceous glands.

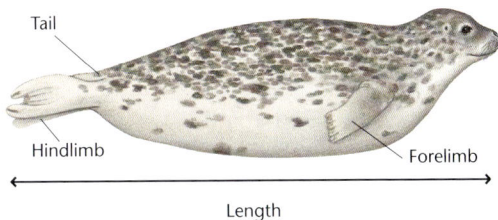

Seal anatomical terms.

Tail

Hindlimb

Forelimb

Length

The seals also make use of sounds which lie within the frequency range audible to the human ear. They are used as signals in ways that are not unlike the behaviour of a number of land mammals. Whether seals can also use echolocation is a matter of debate.

reach a maximum depth of 100 m and remain submerged for up to 13 minutes.

During long, deep dives, seals are unable to renew the oxygen supply to the organs. Part of the problem is resolved by the fact that all but a few organs, such as the brain, can perform oxygen-free (anaerobic) metabolism.

Living areas and migrations

Most seal species live in temperate or cold seas. An exception is the monk seals, a single species of which, the Mediterranean Monk Seal (*Monachus monachus*), is found in the southeastern part of the North Atlantic and in the Mediterranean. A close relative, the Caribbean Monk Seal (*Monachus tropicalis*), is thought to have become extinct in the 1950s.

Migratory pattern varies from species to species. In some it is diffuse, i.e. it occurs over short distances or includes only a part of the population. Other species undertake pronounced long-distance migrations which involve many animals. In addition, stragglers frequently occur outside the normal ranges. Many of the arctic seal species have thus been found farther south. In 1987, for example, a large invasion of Harp Seals took place in the North Sea and inner Danish waters. Bearded Seal and Hooded Seal have also been recorded farther south, on Portuguese coasts and the German east coast, respectively.

Diving behaviour

The dive times and the maximum diving depths of seals vary somewhat from one species to another. The Common Seal can descend to a depth of 500 m, the Harp Seal to 350 m, and the Grey Seal to almost 300 m. The maximum duration of dive varies from 15 to 30 minutes. The Walrus can

Feeding

Most seals are fish-eaters, but crustaceans, cephalopods and other molluscs are also part of the diet.

Seal research

Research on seals seeks answers to largely the same questions as those investigated by whale research. How can species and populations be distinguished from one another? How large are the populations? What are the temporal and spatial distributions of the animals? Do they migrate? How deep and for how long can they dive? When and how do they undertake breeding? What do they eat?

As with the cetaceans, the answers to these questions are based on observations of live animals and examination of dead ones. Seals can be fitted with transmitters, which make it possible to monitor their journeys, and with special devices which record the animals' dive times etc.

Sealing

Seals have been hunted in many places in the North Atlantic ever since the Stone Age. Local exploitation of seals still takes place in Greenland and Canada.

Commercial hunting has also taken place. This was directed first of all towards Walrus, Harp Seal and Hooded Seal in the

Freeze-marking of seals.

arctic regions. In addition, bounties have been offered in many places in the North Atlantic in order to reduce seal populations, as seals are often regarded as pests because they compete with fishermen for fish stocks.

Unintentional catches in drift nets and bottom nets are recorded for many seal species.

Seal-watching

Seal observations are often included as an extra element in many whale tours. Thus, in Greenland, in the right places, Walrus, Bearded Seal, Narwhal and Beluga can be seen on the same trip.

Two different seal heads, Common Seal (left) and Grey Seal (right). Note e.g. the great difference in the way in which the snout is set off from the rest of the head, the size and shape of the eyes, the position of the ear in relation to the eye, and the shape and position of the nostrils.

'Banana posture' of Common Seal at rest.

Field identification

A correct species identification both of swimming and resting seals and of dead ones is desirable. Even though, in principle, the same characteristics are used in each of the three situations, it is more difficult to identify swimming seals. In many places, however, the animals can be observed on land and, if they are approached closely enough, it is possible to study at leisure those characteristics which are used also to identify freshly dead seals.

Coloration

The coloration of seals is determined by the coat, and in only a few cases, such as the Walrus, by the skin colour. In many species, the colour and pattern of the coat change with age. The young of some species are born in a white *lanugo* coat, whereas other species moult into a juvenile coat at the foetal stage. Some species forgo the juvenile coat and moult directly into an adult coat.

Within a species, coat colour can vary both regionally and sexually.

Sexing

When the two sexes do not differ much from each other in size and cannot be distinguished in other ways, sex can easily be determined if the animal is lying on its side (*see* figure). Males have the genital opening in the middle of the belly, whereas that of females is at the rear.

Sexing of seals.

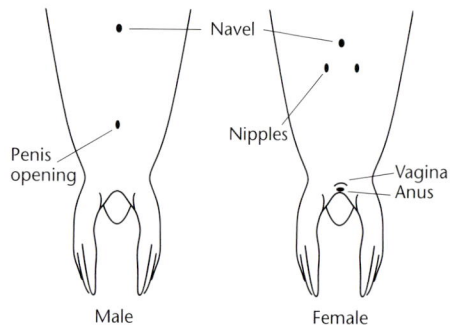

Common Seal.

True seals

This family (Phocidae) comprises species which lack external ears, and which have backward-directed hindlimbs that do not support the body. On the basis of the teeth and cranial characters, two subfamilies are traditionally distinguished: Phocinae (northern seals) and Monachinae (monk seals and Southern Hemisphere seals).

Of the seals discussed here, Common Seal, Ringed Seal, Harp Seal, Grey Seal, Bearded Seal and Hooded Seal belong to the Phocinae, and Mediterranean Monk Seal to the Monachinae.

Group of Common Seals.

Common Seal
Phoca vitulina

ALTERNATIVE ENGLISH NAME Harbor Seal (America)
FRENCH Phoque veau marin
GERMAN Seehund
SPANISH Foca común

Description: Small seal with rounded head. Nostrils form a V when the animal is seen head-on (*see* figure p. 151). The eyes are equidistant between the ear and the nose. Two nipples.
Size: Adults are 1.2–2 m long and weigh 65–140 kg; pups are c. 80 cm long and weigh c. 10 kg.

Common Seal.

Teeth: Three front teeth, one eye tooth and five cheek teeth in each half of upper jaw, and two front teeth, one eye tooth and five cheek teeth in each half of lower jaw, giving the following formula: I3/2, C1/1, PC5/5.
Coloration: The coat is typically spotted.
Variation: Males are a little bigger than females. Coat colour varies both regionally and with age. There are two main colour variants: pale and dark. Most pups are born with a short adult coat.

Behaviour: The seals are seen best at the haul-out sites, which are on sandbanks or deserted stretches of beach. In such places large gatherings of several hundred animals can occur, though there is never close body contact between them. The characteristic haul-out posture is described as 'banana-like', with the head and tail raised and the belly on the ground. Aggressive interactions here seem to be restricted to nipping, pushing and barking. The seals haul out for longer periods during their moult in July and August.

For the most part, only the head shows when the seal is in the water. It can dive to 450 m and remain submerged for up to half an hour.

The Common Seal is a quiet animal. It occasionally produces sounds described as growls and grunts, and sometimes barks. The pups emit a wailing and whimpering sound, although the mother recognises her newborn pup predominantly by smell.

In several places, e.g. on the USA's east coast, in the North Sea and at a few sites in inner Danish waters, the Common Seal shares haul-out sites with the Grey Seal.

Reproduction: The Common Seal becomes sexually mature at an age of 3–6 years. Mating takes place in the water, between February and October. Gestation, including delayed implantation, lasts 11–12 months, and as a rule only a single pup is born. Births generally take place in the period April to July, but may occur some months earlier in more southerly regions. The lactation period lasts 25–30 days.

Food: Fish, squid and crustaceans.

Natural enemies: Killer Whales, large sharks and Polar Bears.

Relationship with man: Common Seals have been hunted since the Stone Age. In the 1800s, they were looked upon as fish thieves and in several countries, including the UK and Denmark, bounties were paid for seals that had been killed.

Until the early 1970s, commercial exploitation took place in the Shetland Islands. In Norway over 1,000 seals were slaughtered in the period 1980-86, and in Iceland more than 117,000 seal pups and 7,300 adults were killed between 1962 and 1987. Today, limited hunting occurs in west Greenland.

Common Seals have been taken into captivity in several places.

Similar species: The Common Seal can be confused with other seal species of the same size, but some of those can be excluded on grounds of range. It can be distinguished from the *Grey Seal*, with which it shares haul-out sites in some places, by its head shape and the position of the nostrils.

Dead Common Seals can be identified by the position and shape of the cheek teeth: they are offset to the side, do not form a straight line, and appear bigger and enlarged compared with those of other seals.

Distribution: Widely distributed in the North Atlantic. In the eastern sectors it is found from Iceland, the Faeroes and Svalbard south to the Irish Sea and Bay of Biscay. A few vagrants have been observed as far south as the Portuguese coast. Large populations live in the North Sea, especially in the Wadden Sea, Skagerrak, Kattegat and Danish waters, and in the Baltic.

In the northwest Atlantic, occurs occasionally in west Greenland and is also rather rare on the east Greenland coast. In contrast, it is common on the Canadian and USA coast south to Massachusetts, with sightings as far south as Florida.

In some places Common Seals are found up rivers. In Canada, there are populations which live throughout the year in fresh water.

Ringed Seal

Ringed Seal
Phoca hispida

FRENCH Phoque annelé
GERMAN Ringelrobbe
SPANISH Foca marbreada

Description: Small, thickset seal (maximum girth can equal 80% of the animal's length) with relatively small, rounded head. The muzzle is short and broad, and the whiskers white. The eyes are close-set. The flippers are small and have pointed ends. Two nipples.

Size: Adults grow up to 1.65 m long and weigh 50–100 kg; pups are on average 60–65 cm long and weigh 4–5 kg.

Teeth: Three front teeth, one eye tooth and five cheek teeth in each half of upper jaw, and two front teeth, one eye tooth and five cheek teeth in each half of lower jaw, giving the following dental formula: I3/2, C1/1, PC5/5.

Coloration: Characteristic, with large white rings on adults.

Variation: Pups are born with a lanugo coat, which they begin to shed after 2–3 weeks; by 6–8 weeks it has fully moulted. The juvenile coat is almost unmarked, silver-coloured on the belly and dark grey on the back. The rings are acquired with age; they may coalesce on some adults, and

Distribution: Ringed Seal has a circumpolar arctic distribution. In the North Atlantic, it occurs in west Greenland, on Labrador coasts, in east Greenland, at Svalbard and in the White Sea. In addition, there is an isolated population in the Baltic Sea, which also contains discrete populations in Lake Ladoga (Russia) and Lake Saimää (Finland). Vagrants may be found farther south.

these individuals can appear almost pale. The populations in Lake Ladoga, in Russia, and Lake Saimää, in Finland, are considered separate subspecies.

Behaviour: Births and lactation take place in special hollows, or lairs, in the ice. There are several types of lair; the most common is oblong, 2.5×3.5 m, with a 30-cm-wide opening at the top. Multi-chambered lairs also exist, which are used not only for young but also as a general resting place. Sometimes the seals also rest in places where the ice is broken up.

Young seals travel much more than do adults, which can be almost sedentary. The longest dive time recorded was of 18 minutes. Does not normally dive very deep, but can probably reach a few hundred metres.

The moult occurs mainly in June and July. The presence of bite marks suggests some rivalry among males. This seal's voice is described as a 'bark' and is heard both day and night throughout the year.

Reproduction: Males and females become sexually mature at a length of 1.3 m, females at an age of 4 years and males at 7 years. Matings take place in spring, and gestation lasts c. 11 months including delayed implantation of 3 months. The pups are born in March and April, but earlier in the Baltic Sea. The lactation period lasts 5–7 weeks.

Food: Small crustaceans (amphipods), fish and squid.

Natural enemies: Killer Whales and Polar Bear. Possibly Walrus.

Relationship with man: Arctic peoples hunt or have hunted the Ringed Seal. In Greenland, between 50,000 and 70,000 animals are killed annually. In the Baltic, large-scale hunting of Ringed Seals has been of a magnitude of several hundred thousand animals. Hunting at its present level is not thought to be a threat to the populations.

Similar species: The greatest risk of confusion is with the *Common Seal*. The Ringed Seal, however, can almost always be identified by its distinct rings, its smaller size and its more thickset build. The cheek teeth are smaller than the Common Seal's and form a straight line.

Ringed Seal.

Harp Seal.

Harp Seal
Phoca groenlandica

FRENCH Phoque du Groenland
GERMAN Sattelrobbe
SPANISH Foca de Groenlandia

Description: Medium-sized seal with slightly pointed head. Two nipples.
Size: Adults are 1.6–1.9 m long and weigh up to 135 kg; newborn pups are c. 90 cm long and weigh 10 kg.
Teeth: Three front teeth, one eye tooth and five cheek teeth in each half of upper jaw, and two front teeth, one eye tooth and five cheek teeth in each half of lower jaw, giving the customary dental formula: I3/2, C1/1, PC5/5.
Coloration: Coat colour changes with age, and there are particular names for the different age classes. Newborn pups ('white coats') have a lanugo coat for 12 days, after which a grey pelt develops ('grey coats'), this being shed after 21 days. Juveniles have a grey coat with small black spots. The adult's coat pattern is very

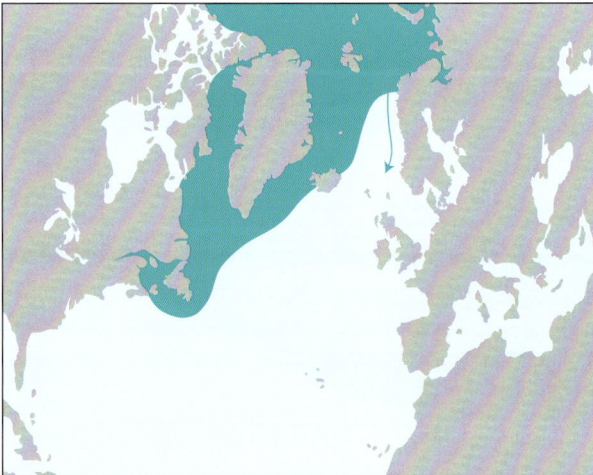

Distribution: The preferred habitat is arctic seas with pack ice. Three populations exist: one in the White Sea, one at Jan Mayen and one in Newfoundland. Stray individuals have, however, been found farther south.

complex: overall, it comprises a black head and a black saddle on the back, while the rest of the coat is whitish to creamy. *Variation:* Males are slightly bigger than females.

Behaviour: Spends most of its time on open sea in groups, but is also seen at the edge of the ice. Dives are generally only brief, but can undertake dives of longer duration down to c. 200 m.

Migrations follow the ice edge and take place between the breeding areas near the southern edge of the ice, where the seals reside in late winter, and the summer quarters farther north. Males assemble in the water around the ice floes, where the young are suckled, and perform the mating display, which includes underwater calls, flipper movements and the uttering of bubbling sounds. Vocalisations are heard only in the mating season.

Moulting starts at the beginning of April; adult males and juveniles begin first, followed by the females.

Reproduction: Becomes sexually mature at an age of 4–7 years. Breeds on pack ice. Matings take place from mid-March to late March, when the young has been weaned. Births take place in February and March, after a gestation of 11–12 months. The lactation period lasts 12 days.

Food: Small and larger crustaceans and pelagic fish.

Natural enemies: Killer Whales and Polar Bear. Possibly Walrus.

Relationship with man: Local hunting on the North American, Greenland and north European coasts has probably been carried out since the Stone Age. Commercial catching of 'white coats' took place at Jan Mayen in the 1800s and continues to this day in Newfoundland.

Similar species: Young animals can be confused with the young of other seal species, but normally the characteristic colour pattern will immediately lead to a correct identification. In the water, Harp Seals in groups can be mistaken for small *cetaceans*, but on closer inspection the seals' body shape will enable clear identification.

Dead, decomposed Harp Seals can be identified only by skeletal characteristics.

Harp Seal with pup.

Grey Seal.

Grey Seal
Halichoerus grypus

FRENCH Phoque gris
GERMAN Kegelrobbe
SPANISH Foca gris

Description: Medium-sized, robust seal with rectangular horse-like head. The nostrils are widely separated. Two nipples.
Size: Adults are up to 2.3 m long and weigh up to 300 kg; newborn pups are c. 1 m long and weigh 11–20 kg.
Teeth: Three front teeth, one eye tooth and five or six cheek teeth in each half of upper jaw, and two front teeth, one eye tooth and five cheek teeth in each half of lower jaw,

giving the following dental formula: I3/2, C1/1, PC5–6/5.
Coloration: Most Grey Seals have a coat that is dark grey on the back and light grey on the belly. On this ground colour there is an irregular contrasting pattern of spots or blotches.
Variation: Males are much bigger and heavier (300 kg) than females (maximum 180 kg) and have a proportionately bigger, broader head. They are also darker than females. Otherwise, coat colour varies quite widely, both individually and geographically. Males become darker with age. Newborn pups have a lanugo coat, which is

Distribution: There are three populations: one in the northern Baltic, one in the northeast Atlantic and one in the northwest Atlantic. The Baltic Sea population previously occurred also in the western Baltic, and in recent years small numbers have begun to re-establish themselves here. The northeast Atlantic population is concentrated around the British Isles, Normandy and the coasts of Norway and Iceland, while the northwest Atlantic population is distributed from coastal Labrador in the north to Cape Hatteras in the south.

replaced over 2–4 weeks by a pale 'adult coat' that has more discrete spots or none at all.

Behaviour: Grey Seals are polygynous, i.e. the males mate with several females and fight over them. They do not, however, defend territories or create harems.

Outside the mating season, Grey Seals disperse over wide areas of sea. They re-assemble to moult and can then be seen on banks and rocks. The moult takes place from March to June.

In the water, it often adopts a 'bottling posture', with the head and upper body sticking out of the water. As a rule, dives for 5 minutes at a time.

Shares haul-out sites with the Common Seal in places where the two species' ranges overlap.

Reproduction: Females become sexually mature when 3–5 years old, males at an age of 4–8 years. Matings (and births) take place from late September to early March. Grey Seals in the British Isles are earliest to start, followed by Norwegian and Iceland ones, then Canadian, and finally the Baltic Sea Grey Seals. In the Baltic population, breeding is thus five months later than in the nearest Atlantic populations. Gestation lasts 11–12 months, and lactation 16–17 days.

Grey Seal.

Food: Fish, squid and crustaceans. Very occasionally, takes seabirds.

Natural enemies: Presumably Killer Whales and large sharks, especially in the case of young Grey Seals.

Relationship with man: Has been subjected to local hunting throughout its range.

Similar species: Small Grey Seals can be confused with other smaller seals. Head shape is, however, a good field character. Dead Grey Seals can be identified by the teeth: only single tips of these project from the gums. The nostrils are almost parallel (*see* figure page 151).

Group of Grey Seals.

Bearded Seal.

Bearded Seal
Erignathus barbatus

FRENCH Phoque barbu
GERMAN Bartrobbe
SPANISH Foca barbuda

Description: Large seal with very small head and comparatively short foreflippers, which make the seal appear longer that it really is. The eyes are small and close-set. The whiskers are strong, relatively long and pale in colour, and are a reasonably good field character. The flippers are peculiar in that the individual fingers are all of the same length. Four nipples, which can be retracted.

Size: Adults are up to 2.5 m long and 360 kg in weight; at birth, the pups are c. 1.3 m long and weigh c. 35 kg.

Teeth: Three front teeth, one eye tooth and five cheek teeth in each half of upper jaw, and two front teeth, one eye tooth and five cheek teeth in each half of lower jaw, giving the following dental formula: I3/2, C1/1, PC5/5.

Bearded Seal with nostrils splayed.

Breathing hole of Bearded Seal.

Coloration: The coat colour of adults is a shade darker on the back than on the underside. Juveniles have a long dark, rippled coat with up to 4 pale transverse bands. The lanugo coat is moulted before birth.
Variation: Females are slightly bigger and heavier than males.

Behaviour: Bearded Seals are solitary, and only rarely are several seen together. The 'bottling posture' is common when resting in the water. The species is quite shy and extremely wary. It always lies with its head close to the water, so that, if disturbed, it can launch off and escape. The young enter the water only a few days after birth.

A northward migration takes place after mating, coinciding with the retreat of the ice. Maintains breathing holes in the ice with its powerful claws.

Reproduction: Males become sexually mature when 6–7 years old, females at 3 years or older. Matings take place in April–May. The gestation period is 11 months when the delayed implantation of c. 2 months is included. Births take place on the pack ice from mid-March to the beginning of May. Lactation lasts 12–18 days.

Distribution: Drift and pack ice at shallower depths in arctic waters south of 80°N. In the North Atlantic, the Bearded Seal is found mainly in Labrador, Greenland, Jan Mayen and Svalbard. It is also recorded in the subarctic zone, in the Gulf of St Lawrence. Stray individuals, however, have been found even farther south, e.g. in the Baltic Sea and, in extreme cases, on the Portuguese coast.

Bearded Seal.

Food: Invertebrates, perhaps supplemented with fish.

Natural enemies: Killer Whale, Polar Bear and possibly Walrus.

Relationship with man: Local hunting of Bearded Seals has taken place throughout the North Atlantic Arctic from earliest times. Commercial hunting has been carried out in the White Sea and the Barents Sea, with a reported kill of a few thousand animals per year. Today, fewer than 900 individuals are caught annually in Greenland. Hunting still continues, but is on a much smaller scale.

Similar species: With its rectangular body and its long whiskers, the Bearded Seal is normally not easily confused with other seals. Unlike other northern seals, it has four nipples. On the ice, however, it can be taken for a young *Hooded Seal*, and in the water confusion with young *Harp Seal* is possible.

Male Hooded Seal.

Hooded Seal
Cystophora cristata

FRENCH Phoque à crête
GERMAN Klappmütze
SPANISH Foca capuchina

Description: Robust seal. The flippers are relatively short. The whiskers are short; on young animals they are dark, whereas they are light-coloured on adults.
Size: Males grow to 2.6 m long and can weigh up to 400 kg, females being smaller; at birth, the pup is 87–115 cm long and weighs 20–30 kg.
Teeth: Two front teeth, one eye tooth and five cheek teeth in each half of upper jaw, and one front tooth, one eye tooth and five cheek teeth in each half of lower jaw,

giving the following dental formula: I2/1, C1/1, PC5/5.
Coloration: The coat colour is silver-grey with irregular black spots or patches. The muzzle is black on adults.
Variation: The male's 'hood', which can, when inflated, double the apparent size of the head, is used for displaying in association with mating contests. By closing the left nostril, the male can also blow a reddish balloon out of it. On juveniles, known as 'bluebacks', the back is blue-grey and the belly whitish; the juvenile retains this coat until its second summer.

Behaviour: Occurs in groups in the mating period, but otherwise often travels

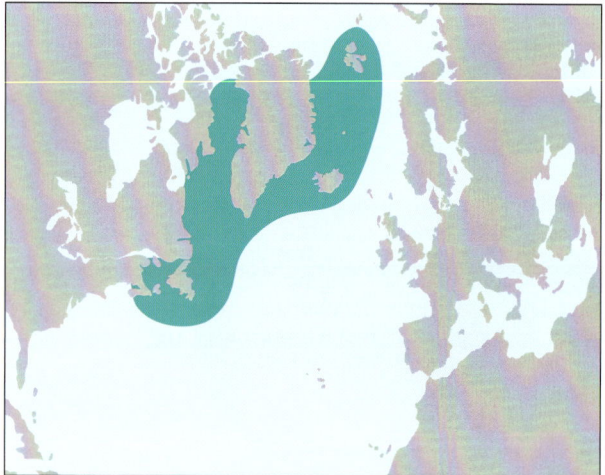

Distribution: Arctic seas with pack ice. Labrador, Davis Strait, Greenland, Jan Mayen and Svalbard. Stray individuals, however, have been recorded farther south, e.g. in Florida, Denmark and Portugal.

solitarily. Dives for up to 15 minutes, often in deep water.

The males hold territories, which they patrol. Rival males fight and can inflict severe wounds on each other. When the Hooded Seal shakes the balloon, it simultaneously emits a high ringing sound. Often lies in loose groups of three, consisting of one male and two females. The females normally maintain a good distance between each other, and are fierce in defence of their young.

The moult of the coat takes place from June to August.

Reproduction: There are two populations with four breeding areas: the Greenland population, which breeds on the west Greenland ice and at Jan Mayen, and the northwest Atlantic population, which breeds at Newfoundland and in the Gulf of Saint Lawrence.

Females become sexually mature at 3–5 years, males at 4–6 years. Mating takes place in the water in April. Gestation lasts 11–12 months, including a delayed-implantation period of 3 months. Births take place in March of the following year on the fast ice, a good distance from the pack ice. Pups are fully weaned after only 4 days, the shortest lactation period of any mammal.

Food: Presumably fish and squid.

Natural enemies: Probably Killer Whales and Polar Bear, and possibly Walrus.

Relationship with man: Hooded Seals have been hunted by man since the Stone Age, and are still killed in Greenland and Canada. Commercial exploitation of their oil and pelts has also taken place since the 1800s to the present day. Up to 83,000 animals were killed annually in Jan Mayen and Greenland between 1946 and 1971, since when the number taken has fallen dramatically. At Newfoundland, annual catches were of the order of c. 15,000 animals, but here, too, fewer have been taken in recent years. Accidental by-catches also occur.

Similar species: Hooded Seals, especially young animals, can be confused with the four other seal species (Common, Ringed, Harp and Bearded Seals) which occur in arctic waters, but coat colour and head shape are normally good field characters. On ice, confusion may occur with the *Bearded Seal*, but that is bigger and has a smaller head. In the water, confusion is most likely with the *Harp Seal*.

Young Hooded Seal.

Mediterranean Monk Seal.

Mediterranean Monk Seal
Monachus monachus

FRENCH Phoque moine de la Méditerranée
GERMAN Mittelmeermönchsrobbe
SPANISH Foca monje del Mediterraneo

Description: Medium-sized seal with comparatively short foreflippers and a tubular body. The head is rather flat and the eyes are widely separated. The muzzle is broad, and has large whisker pads. The nostrils are turned upwards, not forwards as on other true seals, and the whiskers are smooth, not coarse as on other true seals. Four nipples.
Size: Adults are up to 2.8 m long and weigh 250–400 kg; newborn pups are 80–120 cm long and 15–26 kg in weight.

Teeth: Two front teeth, one eye tooth and five cheek teeth in each half of both upper and lower jaws, giving the following dental formula: I2/2, C1/1, PC5/5.
Coloration: Most are dark brown, the belly slightly lighter brown than the back.
Variation: Coat colour varies somewhat between the sexes. The transition from back to belly can be either gradual or sharply demarcated. Some animals have large white areas on the belly. The pups are born in a 'woolly' black coat, which sometimes has white spots; within the first 4–6 weeks this is replaced by a silver-grey juvenile coat. Very little is known about this species' moult pattern and the age-related variations in coat colour.

Distribution: Was formerly widespread in the Mediterranean, but is now reduced to small populations in the Adriatic Sea, the Aegean Sea and the Black Sea and on the north African coast. Occurs also at the Canary Islands, Madeira (Ilhas Desertas) and the West African coast. The majority of the population is today found on the Mauritanian coast (Cape Blanc), where steep coastal slopes ensure undisturbed and safe conditions.

Behaviour: Judging from its behaviour on land, the Mediterranean Monk Seal must be considered one of the least sociable seals. Most of its social contact presumably takes place in the water, where the only hitherto observed mating was also seen.

The species is regarded as sedentary and does not travel widely. Originally it probably lived on sandbanks and open stretches of shore, but nowadays it occurs only in very secluded sites such as caves.

The Mediterranean Monk Seal normally dives to depths of only c. 30 m. As the species is recorded in much deeper water, however, its diving capability is probably far greater.

Reproduction: Females reach sexual maturity at a length of 2.1 m. Estimated age of onset of sexual maturity is 4–6 years for both sexes. Matings (and births) are presumed to occur all year around, with the peak from August to October. Length of gestation is estimated at 11–12 months.

Food: Fish and squid.

Natural enemies: None reported, but possibly Killer Whales. Large shark species may take young animals.

Relationship with man: The seals have probably been exploited in the Mediterranean ever since the first humans began to fish and to exploit marine resources. There are no details of organised, commercial hunting, and the species' marked decline is due perhaps to more indirect causes such as pollution and damage to habitats, overfishing and disturbance.

The species has been kept in captivity in several Mediterranean countries.

Similar species: As the Mediterranean Monk Seal's range does not normally overlap with that of other seal species, there should be no difficulty in identifying it. In the case of dead animals, the position of the nostrils, the colour and nature of the whiskers and, on females, the presence of four nipples are reliable features.

Walruses

The Walrus family (Odobenidae) consists of just a single extant species. It lacks external ears, as do the true seals, but, unlike the latter, it is able to support its body by using the hindflippers. The Walrus does not have a thick coat, but a sparse covering of hair. The skin is 'bark-like', with numerous furrows.

Walrus.

Walrus
Odobenus rosmarus

FRENCH Morse
GERMAN Walroß
SPANISH Morsa

Description: The tusks, up to 50 cm in length, render the Walrus unmistakable. They are already 10 cm long on 2-year-old animals, while 5-year-olds have tusks c. 30 cm long. The foreflippers are equipped with claws which enable the animal to hold on to smooth surfaces.
Size: Grows to up to 3 m long and can weigh 1,000 kg; at birth, the pup is 1–1.2 m long and weighs 45–75 kg.
Teeth: One front tooth, one eye tooth and three cheek teeth in each half of upper jaw, and only one eye tooth and three cheek teeth in each half of lower jaw, giving the following dental formula: I1/0, C1/1, M3/3.

Distribution: Predominantly arctic in drift-ice regions and at edge of ice. Chiefly coastal seas less than 100 m in depth.
Important areas are the White Sea and the Siberian north coast, Svalbard (four localities: southern Edgeøya–Tusenøya, northeast Nordaustland–White Island, Nordporten–Hinlopenstretet, and Moffen Island, now a nature reserve), and also Victoria Island and Franz Josef Island, the east Greenland coast between 70° and 81°N, Davis Strait, Baffin Bay and the Thule district in north Greenland and west Greenland from 66° to 71°45′N, and the high-arctic Canadian islands.
Vagrants have been observed in the North Sea, and Walruses have been seen as far south as Denmark (most recently in 1999).

Coloration: The Walrus's coat is very modest. The most eye-catching feature is the moustache, which consists of sensory hairs. There are not very many hairs on the rest of the body. The skin is up to 8 cm thick and is chestnut-coloured, sometimes with a reddish tinge.

Variation: Old males sometimes have tubercles or protuberances on the back of the neck. The males' tusks are bigger and sturdier than those of females, and males are also generally 20% longer and 50% heavier than females of equivalent age.

Behaviour: The tusks are used as tools. The animals use them as ice picks when climbing on to floes. Gatherings of several hundred animals can form on the ice and at coasts.

The voice is a roar or bark. Under the water and in association with mating, males utter a bell-like sound. The underwater call of the young can best be described as 'twittering'.

Reproduction: Females become sexually mature at 5–6 years of age, males at 8–9 years. Young males, however, are excluded from the mating process for as long as they are 'undersized', i.e. when they have not yet attained their full size. Prior to mating, from January to March, the females assemble at special sites. Mating itself takes place in the water. Gestation is 'protracted' by 4 months owing to delayed implantation of the fertilised egg. The pup is born in May–June after a total pregnancy of 15 months. The female can produce one pup every other year.

Food: Mussels and bivalves. Large males may sometimes use their tusks to kill seals. Walruses forage on soft sea beds at depths of less than 100 m.

Walrus group 'sunning' itself.

Dominant male in centre of group.

Natural enemies: Polar Bears and Killer Whales.

Relationship with man: Has been heavily sought after for its tusks, especially in the 1800s and 1900s. The species was on the verge of extinction when it was granted total protection in 1952. Since then, there have been signs of a slight growth in the population. The Inuit peoples in Greenland and in arctic Canada are licensed to carry out limited hunting.

Small group of Walruses on drifting ice floe.

OTHER MARINE-MAMMAL GROUPS

Besides cetaceans and seals, the manatees and some carnivores are included among marine mammals.

Manatees

Manatees were classified by early taxonomists as plant-eating cetaceans, but they were soon recognised as an independent order of mammals known as Sirenia, a name inspired by the mermaids of Greek mythology. They are also referred to as sirenians.

There are four extant species, in two families: the manatees (Trichechidae), which consist of three species found in the Atlantic's tropical coastal regions and South American and African rivers; and the dugongs (Dugongidae), with a single extant species (*Dugong dugon*) which is a true marine mammal, with a distribution in the Indian Ocean and western Pacific Ocean. A fifth species, the Steller's Sea Cow (*Hydrodamalis gigas*), was found in the northern Pacific, but became extinct at the beginning of the 1700s.

One species, the West Indian Manatee (*Trichechus manatus*), occurs along the North Atlantic coast. A second species, the West African Manatee (*Trichechus senegalensis*), lives in the River Niger and along the West African coast, and vagrants could reach as far north as the Moroccan coast. It is nevertheless included below, because there is a remote chance of encountering it within the region covered by this book.

The only thing that the sirenians have in common with the cetaceans is that they are wholly adapted to a life in water, mostly fresh and brackish water in the case of the manatees. The sirenians are not closely related to the cetaceans, but are in fact distant relatives of the elephants.

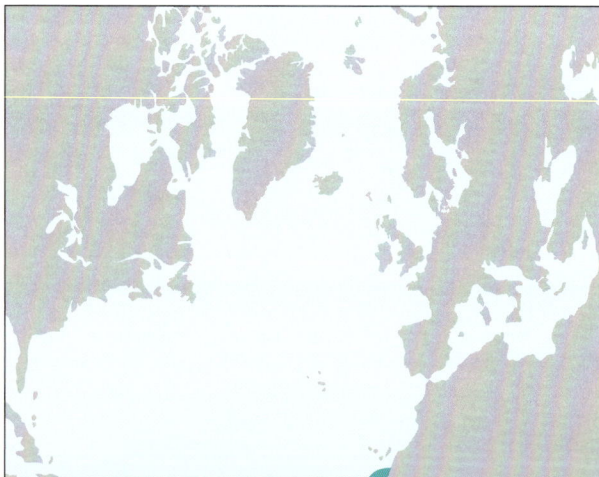

Distribution: Lives in coastal areas, rivers and estuaries. Does not normally move into deep water. The southern coast of Mauritania is normally the species' northern limit, but stray individuals could perhaps be expected farther north.

West African Manatee.

West African Manatee
Trichechus senegalensis

FRENCH Lamatin d'Afrique
GERMAN Westafrikanischer Manat
SPANISH Vaca marina del Africa

Description: Typical manatee, but comparatively slim with short head. The eyes are clearly visible. Paddle-shaped flippers and tail.
Size: Adults are 3–4 m long and weigh 750 kg; newborn calves are c. 1 m long.
Teeth: Has 5–7 functional (cheek) teeth in each dental strip in both upper and lower jaws. As the teeth gradually wear down, they are replaced from the back by new, erupted teeth. Newborn calves have rudimentary teeth which disappear later.
Coloration: Hairless except for stiff bristles on the upper lip. Skin colour is grey-brown and the bristles are white.
Variation: There is no significant size difference between the sexes. Females, however, can be recognised during the lactation period by their conspicuous, swollen mammary glands in the axillary areas.

Behaviour: Solitary, but mothers and young are seen together; aggregations of up to 15 animals are also known.
 Mother and calf communicate vocally, but probably also make use of olfactory and visual senses. Manatees swim slowly, and often float lethargically at the surface. They rarely submerge for long at any one time, and normally keep within a few hundred metres of the coast.

Reproduction: There is no information on age and size at sexual maturity. Is thought to breed all year around, but with increased activity in the rainy season. The young are born in shallow lagoons.

Food: Plants, mangrove leaves and mussels have been found in stomachs of dead animals; may also eat fish.

Natural enemies: None known. It is not unlikely that Killer Whales might kill manatees in coastal waters.

Relationship with man: Hunted by local people for its meat and oil.

Similar species: With its slow and sedate movements, and because of the shape of the head and tail, the West African Manatee cannot be confused with other animals. The *West Indian Manatee* is very similar to it, but is plumper and has a more pointed head. The *Walrus* is similar in being almost hairless, but it inhabits a totally different region and is at once identifiable by its tusks.

West Indian Manatee.

West Indian Manatee
Trichechus manatus

FRENCH Lamatin des Caraïbes
GERMAN Westindischer Manat
SPANISH Vaca marina del Caribe

Description: Quite plump, with paddle-shaped, flexible forelimbs and paddle-shaped tail. The body has a number of folds and wrinkles. Has stiff bristles on the upper lip and small fine hairs on the rest of the body. Two semicircular nostrils. Each flipper has three or four claws.

Size: Adults are up to 3.5 m long and weigh up to 1,600 kg; newborn calves are c. 1.2 m long and weigh c. 30 kg.
Teeth: Has 5–7 teeth in each jaw. At birth there are two front teeth, but these disappear later.
Coloration: Skin colour is grey-brown, sometimes with a greenish cast caused by algal growth. Young are often darker, some even black.
Variation: No significant difference between the sexes.

Distribution: Coastal areas and swamps along the east coast of America, from Florida in the south to Cape Cod in the north.

Behaviour: Very similar to West African Manatee. Mother-young pair and small groups are most frequent, but larger groups of about a dozen can occur. The manatees communicate by sound, but probably also make use of olfactory and visual cues. Swims slowly, and often lies lethargically at the surface. Rarely submerges for long at any one time, and generally keeps within a few hundred metres of the coast.

Reproduction: Very little is known. May take place throughout the year, but peaks in spring and summer. Gestation is thought to last 12 months.

Food: Aquatic plants such as water hyacinths and seagrasses. In some regions, the species also eats parts of the mangrove vegetation and invertebrates such as tunicates (sea squirts), as well as fish.

Natural enemies: None known.

Relationship with man: The population has been considerably reduced by hunting, and is now threatened. Many individuals die as a result of collisions with fast-moving boats of various kinds.

Similar species: Behaviour and appearance should render the species unmistakable. The only other sirenian in the North Atlantic, the *West African Manatee*, lives on the other side of the ocean and is not known to undertake transoceanic migrations. Stray individuals of the *Walrus*, which is superficially similar to the manatee, could theoretically reach the manatee's range along the American east coast. This applies especially to young Walruses, which lack large tusks, but on closer inspection the manatee is unmistakable. The behaviour is also totally different.

Marine carnivores

Apart from the seals, which are sometimes placed in the order of carnivores (Carnivora), three species of carnivore are regarded as marine mammals. These are the Sea Otter (*Enhydra lutris*), which lives in the north Pacific Ocean, the South American Marine Otter (*Lutra felina*) and the Polar Bear (*Ursus maritimus*), the last being the only one to occur in the North Atlantic. Two freshwater otters, however, have coastal populations in the region covered by this book (*see* page 180).

The Polar Bear represents the most recent 'return' to the sea among the mammals. It is a 'Brown Bear' that has been able to adapt to arctic waters and ice conditions, as it possesses a layer of fat (blubber) and a thick coat. The teeth have become adapted to a largely carnivorous diet, whereas the Brown Bear's teeth are suited to a more varied diet.

Polar Bear
Ursus maritimus

FRENCH Ours blanc
GERMAN Eisbär
SPANISH Oso polar

Description: With its body shape, its rectangular head and its white coat, the Polar Bear is unmistakable. The big oar-like front paws are adapted for swimming.
Size: Adults are on average 2–2.5 m long from tip of nose to tip of tail and weigh 300–800 kg; cubs are very small at birth, with a weight of c. 1 kg, and they grow to 15 kg before leaving the maternal snow den.
Teeth: Has 38–42 teeth, as expressed in the following formula: I3/3, C1/1, PM2-4/2–4, M2/3, where PM stands for premolars and M for molars (together termed PC).
Coloration: The coat is normally whitish, and the nose, the lips and the feet black.
Variation: Males are bigger than females. Coat colour can vary and show elements of yellow, grey or even brown.

Behaviour: Apart from females with their young, Polar Bears are normally seen only singly, though large corpses of e.g. a whale can attract a larger number. The Polar Bear is active throughout the year. Pregnant females, however, excavate breeding dens, where they remain until the cubs are big enough to leave them. The cubs accompany the female until they are 2½ years old. Males may also decide to dig temporary dens during severe weather conditions.

Polar Bears are known to have swum over 100 km across open sea, and they migrate over impressive distances. For instance, bears marked at Svalbard have been found again in southwest Greenland (a distance of over 3,000 km). They can run at 25–30 km/h. It has been shown that Polar Bears generally move on average 40 km per day. The Polar Bear is also an indefatigable swimmer, and it can remain for several hours in the cold water. It also dives at times, but this involves only short superficial dives.

Reproduction: Males become sexually mature when 3 years old; females undergo their first pregnancy when they are 6 years old. Matings take place from late March to mid-May. Implantation is thought to be delayed, so the total gestation amounts to between 195 and 265

Polar Bear.

days. One, two or exceptionally three cubs are produced per litter; they are naked and blind at birth.

Food: Seals, Belugas, Narwhals, fish and sometimes seabirds. On land and in summer, it eats grass and berries. Takes carrion at times.

Natural enemies: None.

Relationship with man: Has been hunted for thousands of years. Until the 1970s a couple of hundred animals were killed annually. The population had by then diminished so greatly that an international conservation agreement was signed in 1973. At Svalbard, the Polar Bear may now be killed only in self-defence. Between 100 and 125 Polar Bears are killed yearly in Greenland. Kept in numerous zoos.

Distribution: Occurs throughout the arctic region north to at least 84°N, but has now disappeared from the Gulf of St Lawrence and Newfoundland. Found in northern Canada and Greenland. Records in west Greenland relate to animals which have travelled around Kap Farvel with the sea ice. Occurs also in east Greenland, Jan Mayen and Svalbard. Has also been reported from the White Sea and north Russia.

Sea and coastal otters

The world's 13 species of otter constitute a subfamily of the mustelids (Mustelidae) characterised by the possession of webs between the toes of the fore- and hindlimbs. The majority of otter species are associated with fresh water, but two species live almost exclusively in the sea and are therefore traditionally regarded as marine mammals. They are the Sea Otter (*Enhydra lutris*) and the South American Marine Otter (*Lutra felina*).

In the North Atlantic there are no 'true' sea otters, but two species, the Eurasian Otter (*Lutra lutra*) and the American River Otter (*Lontra canadensis*), contain populations that exploit the sea's resources in some places, especially in regions with rocky coasts and limited access to fresh water. Nevertheless, they are still dependent on both land and fresh water. These two otter species are so similar to one another in both appearance and lifestyle that they may be regarded as ecological parallels. The American species, however, has a more rounded head and its nasal region (rhinarium) is bigger. The details given below concern only those aspects which are relevant to the coastal populations.

Lutra lutra

Lontra canadensis

Distribution: Otters were formerly found along all European coasts except Iceland, and also on the north African coast, and can potentially be seen in all those places where otter populations are still found. Coastal populations are described from western Scotland, Shetland, the Orkneys and Ireland, and from Norway north of Trondheim.

The American River Otter is found along the American east coast from the Labrador peninsula in the north to Florida in the south. Coastal populations exist in Newfoundland.

Eurasian Otter.

Eurasian Otter
Lutra lutra

FRENCH Loutre commune
GERMAN Fischotter
SPANISH Nutria

American River Otter
Lontra canadensis

FRENCH Loutre commune américaine
GERMAN Amerikanischer Fischotter
SPANISH Nutria americana

Description: Mustelid with small ears and distinct large tail. Webs between toes on all four legs. Coat consists of brown guard hairs with a dense underfur.
Size: Adults have a body length of 60–80 cm, and a well-demarcated tail 35–50 cm long; adults weigh 6–12 kg.
Teeth: Has a typical carnivore dentition: I3/3, C1/1, PM4/3, M1/2.
Coloration: Dark brown back and flanks; underside and tail paler, sometimes almost white.
Variation: Males are both bigger and longer than females.

Behaviour: Encounters with otters are most likely to involve single animals or a female with young foraging at the shoreline. Otters are dependent on the presence of fresh water in their habitat for grooming and as drinking water. Where the sea holds an abundance of fish, the otters may confine themselves to just a few kilometres of coast. Normally nocturnal, but on the coast its activity is frequently related to high and low tides and is therefore diurnal.

Reproduction: In the European species, the female gives birth to a litter of 2–5 young after a gestation period of c. 60 days. In the American species, foetal development enters a stage of 'dormancy' and gestation is thus protracted to 10–11 months. The young are blind at birth and have a greyish juvenile coat. Females become sexually mature at 2 years of age, while males need to be more than 2 years old to be able to participate in breeding.

Swimming Eurasian Otter.

Food: The marine diet of otters consists mostly of coastal fish, such as eels, butterfish and blenny, and various crabs.

Natural enemies: Potential enemies are large predators such as the White-tailed Eagle, Polar Bear and Wolf from the land-ward side, and perhaps Killer Whales from the sea.

Relationship with man: Otters have declined massively in many places. This is due primarily to hunting, disturbance and habitat alteration. Deaths from water-borne traffic and capture in fish traps are also documented. Nothing is known about particular conditions which may affect coastal otters.

Similar species: On land, the otter cannot be confused with any other marine mammal. Size and tail length render it unmistakable. In the water, it can perhaps be confused with certain small seals, but the swimming pattern and the clearly visible tail mean that there are usually no problems in arriving at the correct identification.

Glossary

beak: forward-projecting jaws, or snout, of cetacean

bioluminescent: referring to light produced by biochemical means

blow: the visible column of air exhaled by a cetacean

blowhole: the nasal opening of a cetacean

bow-riding: swimming in the pressure wave (bow wave) created by a moving ship or by a large whale

breaching: of a cetacean, leaping clear (or nearly so) of the surface, generally landing with loud splash

callosity: lumpy protuberance on head of right whale

cape: darker area on the back of certain cetaceans

continental slope: steep slope at outer part of shelf (*see* below), extending down to ocean bed

dorsal: referring to the back, or upperside

epipelagic: of the upper region of the sea, down to a depth of c. 100 m

flipper: forelimb of a cetacean or manatee

flipper-slapping: raising a flipper out of the water and slapping it down on to surface

flukes: the horizontal tail of a cetacean

halocline: layer of sea water where salinity undergoes rapid changes (cf. thermocline)

juvenile: young animal independent of mother, but not yet sexually mature

krill: small shrimp-like crustaceans living in oceanic waters

lanugo coat: first fur coat of seal pup, often shed before birth

logging: of whales, lying still (resting) at the surface

melon: the bulging forehead of a toothed whale

pelagic: relating to the open ocean

pod: a co-ordinated group of cetaceans

polyphyletic: of a group of organisms which have evolved from more than one ancestor

shelf (continental shelf): gently sloping seabed extending from low-water mark at coast out to the start of the continental slope (*see* above)

spyhopping: holding the head vertically above water surface to observe surroundings

stranding: the phenomenon of a cetacean coming ashore (dead or alive)

tail-slapping: raising of tail flukes and slapping them down on to surface (also referred to as lob-tailing)

tail stock: rear part of a cetacean's body, between dorsal fin and tail flukes

temperate: descriptive of mild regions at middle latitudes (40–60°), between polar and subtropical waters

thermocline: layer of sea water where temperature undergoes rapid changes (cf. halocline)

tunicates: a subphylum of the Chordata having a leathery body

upwelling: of nutrient-rich waters which rise to the surface from deeper parts of ocean

Bibliography

Berta, A., & Sumich, J. L. 1999. *Marine Mammals: Evolutionary Biology*. Academic Press, London.

Carwardine, M. 1995. *Whales, Dolphins and Porpoises*. Dorling Kindersley Handbooks, Dorling Kindersley, London.

Carwardine, M., Hoyt, E., Fordyce, R. E., & Gill, P. 1998. *Whales and Dolphins: The Ultimate Guide to Marine Mammals*. HarperCollins, London.

Clapham, P. 1997. *Whales*. Colin Baxter World Wildlife Library, UK.

Cresswell, G., & Walker, D. 2001. *Ocean Guides. Whales & Dolphins of the European Atlantic: The Bay of Biscay and the English Channel*. WILDGuides, Old Basing, Hampshire, UK.

Evans, P. G. H. 1987. *The Natural History of Whales and Dolphins*. Christopher Helm, London.

Evans, P. G. H. 1995. *Guide to the Identification of Whales, Dolphins and Porpoises in European Seas*. Scottish Natural Heritage, UK.

Evans, P. G. H., & Raga, J. A. (eds.) 2001. *Marine Mammals: Biology and Conservation*. Kluwer Academic/Plenum Publishers, New York, Boston, Dordrecht, London & Moscow.

Fontaine, P.-H. 1998. *Whales of the North Atlantic. Biology and Ecology*. Editions MultiMondes, Canada.

Gaskin, D. E. 1982. *The Ecology of Whales and Dolphins*. Heinemann, Portsmouth, New Hampshire.

Hoelzel, A. R. (ed.) 2002. *Marine Mammal Biology: an evolutionary approach*. Blackwell Publishing, Oxford, UK.

Jefferson, T., Leatherwood, S., & Webber, M. 1993. *Marine Mammals of the World*. FAO, Rome.

Katona, S. K., Rough, V., & Richardson, D. T. 1993. *A Field Guide to Whales, Porpoises and Seals from Cape Cod to Newfoundland*. Smithsonian Institution Press, USA.

Leatherwood, S., Reeves, R. R, & Foster, L. 1983. *Sierra Club Handbook of Whales and Dolphins*. Sierra Club, USA.

Macdonald, D. (ed.) 2001. *The New Encyclopedia of Mammals*. Oxford University Press, Oxford.

Mann, J., Connor, R. C., Tyack, P. L., & Whitehead, H. (eds.) 2000. *Cetacean Societies, Field Studies of Dolphins and Whales*. University of Chicago Press, London.

Martin, A. R. 1990. *Whales and Dolphins*. Salamander Books Ltd, London.

Payne, R. 1996. *Among Whales*. Bantum Press, USA.

Perrin, W. F., Wursig, B., & Thewissen, J. G. E. (eds.) 2002. *Encyclopedia of Marine Mammals*. Academic Press, London.

Pryor, K., & Norris, K. S. (eds.) 1991. *Dolphin Societies Discoveries and Puzzles*. University of California Press, Berkeley & Los Angeles, CA.

Reeves, R., Stewart, B. S., & Leatherwood, S. 1992. *Sierra Club Handbook of Seals and Sirenians*. Sierra Club, USA.

Reynolds, J. E., & Rommel, S. A. (eds.) 1999. *Biology of Marine Mammals*. Smithsonian Institution Press, USA.

Ridgeway, S. H., & Harrison, R. J. 1981–99. *Handbook of Marine Mammals*. Vols 1–6. Academic Press, London.

Riedman, M. 1990. *The Pinnipeds: seals, sea lions, and walruses*. University of California Press, Berkeley, Los Angeles & London.

Vialle, S. 1997. *Dolphins and Whales from the Azores*. Portugal.

Wilson, B. 1998. *Dolphins*. Colin Baxter Photography Ltd, UK.

Wuirtz, M., & Repetto, N. 1998. *Whales and Dolphins: Guide to the Biology and Behaviour of Cetaceans*. Swan Hill Press, UK.

Useful websites

European Cetacean Society: www.inter.nl.net/users/J.W.Broekema/ecs/
Greenpeace: www.greenpeace.org
Hebridean Whale and Dolphin Trust: www.gn.apc.org/whales
Marine Conservation Society: www.mcsuk.org
Organisation Cetacea (ORCA): www.orcaweb.org
Sea Watch Foundation: www.seawatchfoundation.org.uk
Shetland Marine Mammal Group: www.wildlife.shetland.co.uk
Whale and Dolphin Conservation Society: www.wdcs.org
WhaleNet: www.whale.wheelock.edu
Whale-Watching-Web: www.physics.helsinki.fi/whale

List of photographers

Index